Anna Karnauhova
Meanders

Also of interest

Infinite Ergodic Theory of Numbers
Marc Kesseböhmer, Sara Munday, Bernd Otto Stratmann, 2016
ISBN 978-3-11-043941-0, e-ISBN (PDF) 978-3-11-043942-7,
e-ISBN (EPUB) 978-3-11-043085-1

Ergodic Theory. Advances in Dynamical Systems
Idris Assani, 2016
ISBN 978-3-11-046086-5, e-ISBN (PDF) 978-3-11-046151-0
e-ISBN (EPUB) 978-3-11-046091-9

Positive Dynamical Systems in Discrete Time. Theory, Models, and Applications
Ulrich Krause, 2015
ISBN 978-3-11-036975-5, e-ISBN (PDF) 978-3-11-036569-6,
e-ISBN (EPUB) 978-3-11-039134-3

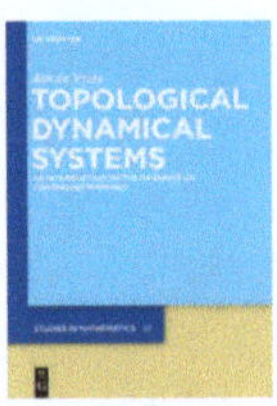

Topological Dynamical Systems. An Introduction to the Dynamics of Continuous Mappings
Jan de Vries, 2014
ISBN 978-3-11-034073-0, e-ISBN (PDF) 978-3-11-034240-6,
e-ISBN (EPUB) 978-3-11-037459-9

Planar Dynamical Systems. Selected Classical Problems
Yirong Liu, Jibin Li, Wentao Huang, 2014
ISBN 978-3-11-029829-1, e-ISBN (PDF) 978-3-11-029836-9,
e-ISBN (EPUB) 978-3-11-038914-2

Anna Karnauhova

Meanders

Sturm Global Attractors, Seaweed Lie Algebras and Classical Yang–Baxter Equation

DE GRUYTER

Mathematics Subject Classification 2010
35-02, 65-02, 65C30, 65C05, 65N35, 65N75, 65N80

Author
Dr. Anna Karnauhova
Institute of Mathematics
Arnimallee 3
14195 Berlin
Germany
anna.karnauhova@fu-berlin.de

Doctoral dissertation at the Free University of Berlin, 2016.
Referees: Prof. Dr. Bernold Fiedler, Prof. Dr. Carlos Rocha.

ISBN 978-3-11-053147-3
e-ISBN (PDF) 978-3-11-053302-6
e-ISBN (EPUB) 978-3-11-053171-8
Set-ISBN 978-3-11-053303-3

Library of Congress Cataloging-in-Publication Data
A CIP catalog record for this book has been applied for at the Library of Congress.

Bibliographic information published by the Deutsche Nationalbibliothek
The Deutsche Nationalbibliothek lists this publication in the Deutsche Nationalbibliografie; detailed bibliographic data are available on the Internet at http://dnb.dnb.de.

Cover image: amoklv/iStock/thinkstock
Printing and binding: CPI books GmbH, Leck
♾ Printed on acid-free paper
Printed in Germany

www.degruyter.com

To my family and friends.

Contents

Preface

Meanders – Historical Background

Originally, *meanders* were introduced by Vladimir I. Arnol'd in [1, §3 Projective Topology of Meanders]: "*A meander shall be understood to be a connected oriented non-self-intersecting curve which intersects a fixed oriented line in several points.*"

Usually, a meander is viewed as a *river*, the fixed oriented line as a *road* and the intersection points as *bridges*. In Figure 1, we have depicted the eight meanders[1] with exactly five vertices which indicate the intersections of the curve with the straight line. For simplicity we represent the resulting arches of the river between bridges by half-circles. If we assign labels to each vertex in the order in which the oriented curve intersects the oriented straight line we get a permutation, called a *meander permutation*, cf. [1] and Figure 1. The meander permutations can be read from the numbers in the black nodes following the direction of the straight line. It is natural to formulate the following question: Given an arbitrary permutation is it a meander permutation?

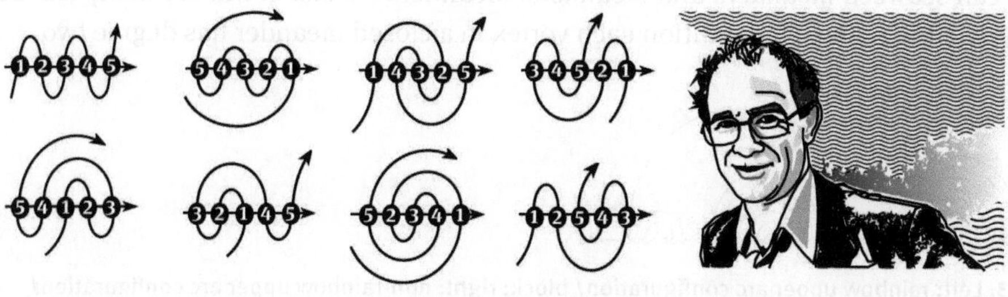

Fig. 1: Vladimir I. Arnol'd (right) and eight possible meanders crossing the oriented straight line five times (left); the numbers inside the black nodes indicate the order in which the straight line was traversed.

Our work will originate from a combinatorial description of meanders. Precisely, there is the following observation we may think of: For a non-self-intersecting oriented curve intersecting the straight line transversely we can view the intersection points as vertices and the segments between them as directed edges.

1 These meanders were originally considered by Arnol'd in [1, p. 721] in order to illustrate the *meander permutations* on five elements.

DOI 10.1515/9783110533026-001

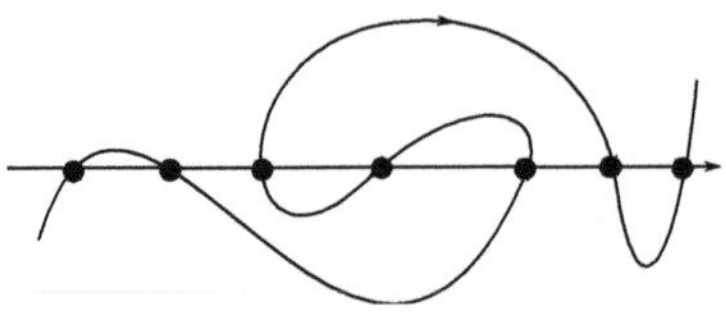

Fig. 2: A non-self-intersecting oriented curve intersecting a fixed oriented line transversely.

Outline and Goals of this Work

One of the main interests of the underlying work is devoted to the study of the connectivity of *meanders* which differ from meanders as defined by Arnol'd. *Meanders* will be introduced as embedded planar graphs. In our study we will not consider meander permutations. Instead we will introduce embedded planar graphs in the Euclidean plane $\mathbb{E} = \mathbb{E}^+ \cup \mathbb{E}^-$ by a sequence of numbers representing the number of unoriented half-circles, i.e., *arcs*. We note, that it would be convenient to introduce the graphs in the Lobatchevsky's plane, where the shortest paths are half-circles. By imposing some additional rules on how to connect the vertices, i.e., by introducing *rainbow* and *non-rainbow* arc configurations/ blocks, see Figure 3, we obtain different graphs which we will call *seaweed meanders* and *meanders*. Meanders consist of finitely many closed curves in the plane. By definition each vertex in a closed meander has degree two.

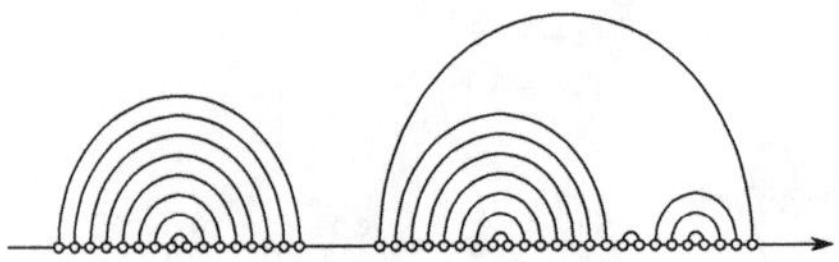

Fig. 3: Left: rainbow upper arc configuration/ block; right: non-rainbow upper arc configuration/ block.

We will introduce the so called *collapse* allowing us to derive results on connectivity. Essentially, the *collapse* may be interpreted in the following way. From topological point of view the collapse is the gluing of two neighboring vertices and arcs.

After the collapse of *closed* meanders, i.e., Jordan curves in the plane, we obtain three different types of sets of *collapsed meanders*. These occurred in the literature in different mathematical disciplines. Different types of collapsed meanders[2] are:

2 Note, that collapsed meanders are not necessarily open since there exist collapsed meanders which entirely consist of cycles. However, there are collapsed meanders which entirely consist of paths. In this case, we consider them to be open.

i) *Morse meanders*, i.e., connected non self-intersecting curves intersecting transversely the horizontal line from south-west to north-east with positive Morse vector; we denote the set of Morse meanders by $\mathscr{M}or^{coll}$,
ii) *collapsed seaweed meanders*, i.e., meanders with collapsed rainbow arc configurations; collapsed seaweed meanders are not necessarily connected; the set of collapsed seaweed meanders is denoted by $\mathscr{S}^{coll}$,
iii) *Frobenius meanders*, which are connected collapsed seaweed meanders; the set of Frobenius meanders is denoted by $\mathscr{F}^{coll}$,
iv) *collapsed meanders*, which generalize collapsed seaweed meanders and inter alia consist of branched curves resulting from the existence of *forks*; the set of collapsed meanders is denoted by $\mathscr{M}^{coll}$.

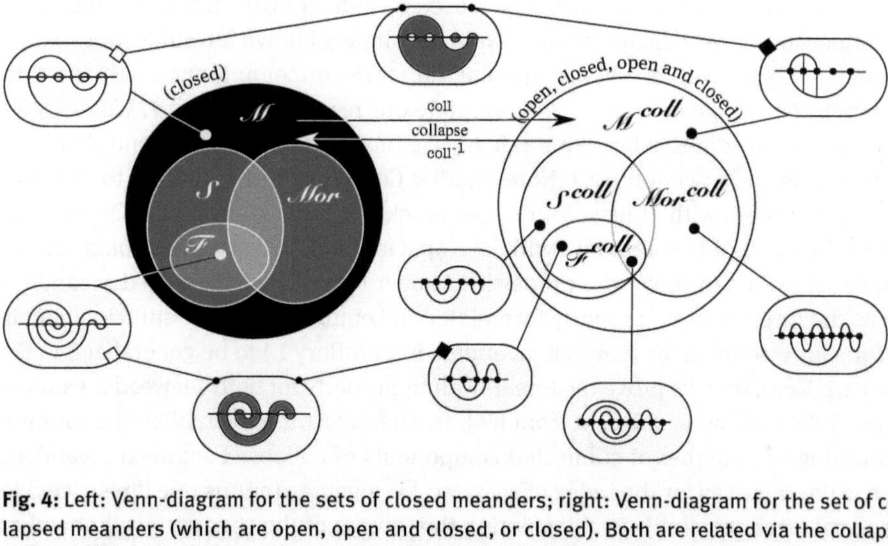

Fig. 4: Left: Venn-diagram for the sets of closed meanders; right: Venn-diagram for the set of collapsed meanders (which are open, open and closed, or closed). Both are related via the collapse indicated by arrows between the two Venn-diagrams. Illustration of the collapse on two examples in the top middle and left lower part of the figure.

The second aspect concerning the importance of the collapse, is summarized by the following Venn-diagrams in Figure 4. On the one hand, Morse meanders are connected collapsed meanders with positive Morse vectors, meaning that all Morse indices are positive. They are closely related to the description of global *Sturm* attractors of semi-linear parabolic partial differential equations with Neumann boundary condition, cf. [11]. On the other hand, collapsed seaweed meanders[3] were introduced in [7] in order to represent the so called seaweed Lie algebras, i.e., Lie algebras preserving two

3 Originally, they were called *meanders* cf. [7]

opposite flags. The main purpose of introducing seaweed meanders was to compute the index of the corresponding seaweed Lie algebra. Collapsed seaweed meanders are not necessarily connected. However, the most interesting ones are connected since they represent seaweed Lie algebras ($\subseteq \mathfrak{gl}_n$) of index one which are called *Frobenius*. Therefore we shall call connected collapsed seaweed meanders *Frobenius meanders*. By a result of Drinfel'd, cf. [2], one can construct a solution to the classical Yang–Baxter equation on Frobenius Lie algebras.

Summary

In Chapter 1, we give a proof of the Conjecture 19 which was formulated in [6] and appeared reformulated as the last sentence in the published paper [5, Ongoing Work]. Precisely, we disprove the existence of a closed formula for the number of connected components of seaweed meanders similar to the already known formulae for seaweed meanders with at most four blocks in total, i.e., three upper and one lower block or two upper and two lower blocks. We formulate our result in Theorem 1. We first give a proof for seaweed meanders with at least five blocks in Section 1.2.1 and then generalize our result in Section 1.2.2. Note, that by Construction 1 it suffices to consider seaweed meanders with at least four upper blocks and one lower block. On our way towards the proof of Theorem 1 we will develop a method, which we call *collapse*. The collapse allows us to prove the properties known for collapsed seaweed meanders, such as the linear-scaling-property formulated in Lemma 1.5 and the sufficient condition for connectedness for seaweed meanders in Corollary 1.1 to be generalized in Corollary 2.2. Note, that we prove the linear-scaling-property for both seaweed meanders and meanders which is different from [24]. In Theorem 1.1, we establish the formula for counting the number of connected components of collapsed seaweed meanders, which is the same[4] as for the index of seaweed Lie algebras (minus one if restricted to the special linear algebra). By considering the inverse of the collapse, we know that the inverse of collapsed seaweed meanders are seaweed meanders. Indeed, it would be rather convenient to represent a seaweed Lie algebra by a seaweed meander rather than by its collapsed version as it was introduced in [7]. The counting of components of seaweed meanders is less involved. For seaweed meanders, we only need to count the number of closed Jordan curves. This aspects will be discussed in Chapter 7 as well as the construction of new *Frobenius meanders* which we obtain from Chapter 4 on Morse meanders. In Chapter 2 we will extend our perspective to arbitrary meanders.

The aim of Chapter 3 is to introduce Morse meanders which are closely related to the description of the global Sturm attractors of semi-linear parabolic partial differential equation. In light of our perspective from the first chapter we will recall the defini-

4 This is true for subalgebras of general linear algebra.

tion of Morse meanders. We will be concerned with the inverse classification problem of Morse meanders. Namely, given a connected meander with negative Morse indices what are possible ways to raise Morse indices in order to get rid of negative ones. This investigation will be useful for getting rid of negative Morse indices of left one-shifted elements to be introduced in Chapter 4.

In Chapter 4 we give a combinatorial description of Morse meanders by introducing the right one-shift map, denoted by $\mathtt{shift}_{[1]}$. By Theorem 4.1 the right one-shift of closed meanders with same configuration of arcs in the upper and lower half-planes are Morse meanders. The left one-shift map $\mathtt{shift}_{[-1]}$, which may be viewed to be dual to the right one-shift map, forces us to recover the negative Morse indices. In order to achieve this, we apply the maps mut and $\mathtt{sus}_{(\cdot)}$ from the previous Chapter 3. By using concatenations, left/right one-shift maps and suspensions we construct Morse meanders of type I, II, III and IV. In comparison to the results of [15], we only construct a subset of pitchforkable shooting curves. However, since our description is combinatorial, it will be possible to study isomorphism questions on connection graphs raised in [9]. The problem of classification of Morse meanders, however, remains open, since the left and right one-shifts of same compositions of upper and lower blocks are not sufficient in order to construct all Morse meanders, which is the topic of the discussion in the end of the chapter under consideration.

The goal of Chapter 5 is set in the definition of connection graphs of type I, II, III and IV in correspondence with the four types of Morse meanders introduced in Chapter 4. We will characterize explicitly these connection graphs and obtain some results for two connection graphs being isomorphic, respectively non-isomorphic. Precisely, we prove in Theorem 5.5 that for two connection graphs of type I for being isomorphic the necessary and sufficient condition is that the corresponding canonical trees differ by reflections on its maximal subtrees. The result in Theorem 5.7 of the last section shows that connection graphs of different types are non-isomorphic.

In the remaining chapters we will relate our results to the already known results which appeared in different scientific contexts, such as the Temperley–Lieb algebra, representations of seaweed Lie algebras, constant solutions of the classical Yang–Baxter equation.

Acknowledgements

We would like to thank Prof. Dr. Bernold Fiedler and Prof. Dr. Carlos Rocha. This book has taken a remarkable advantage of the ideas and many discussions with Vincent Trageser, Dr. habil. Stefan Liebscher, Dr. Juliette Hell and Prof. Dr. Matthias Staudacher.

1 Seaweed Meanders

In this chapter we introduce *seaweed meanders* as embedded planar graphs. The name *seaweed meanders* is related to the work [7] by V. Dergachev and A. Krillov, who introduced meander graphs in order to simplify the index-computation of – as they call them – Lie algebras of *seaweed type* or *seaweed Lie algebras*. By the *collapse*, *meander graphs* as introduced in [7] are *collapsed seaweed meanders*. In Chapter 7, we explain how to assign a seaweed meander to a seaweed Lie algebra rather than its collapsed form as was originally introduced in [7].

Our combinatorial viewpoint set in the first two chapters is important for providing a rigorous mathematical description for *seaweed meanders* and *meanders*. More importantly, it offers a possibility for combinatorial arguments in the investigation of connectivity of both *seaweed meanders* and *meanders*. In Chapters 3 and 4, we will be concerned with *Morse* meanders which play a special role in the description of the *global Sturm attractors* of semi-linear parabolic partial differential equations with Neumann boundary conditions. The notion of collapse will be also essential when considering Morse meanders. The left and right *one-shifts* introduced in Chapter 4, which yield Morse meanders under additional assumptions and requirements are introduced by vertex-gluing which is similar to the idea of *collapse*. The main difference between the two concepts, the *collapse* and the *one-shift*, is that in the first case a pair of arcs is also glued together whereas for the latter only vertices are identified.

As was already mentioned in the introduction, our interest of investigation for both types of *meanders* will be set in the connectivity. The main ingredient in the study of connectivity of *meanders* will be given by the concept of the *collapse* for both *seaweed meanders* and *meanders*. We note, that the collapse as well as the Obstruction-Theorem 1 proven in this chapter is our contribution to the joined work [18] with S. Liebscher.

1.1 Seaweed Meanders

In the sequel, $\mathbb{E}$ denotes the Euclidean plane and $\mathbb{E}^+$ and $\mathbb{E}^-$ the upper and lower Euclidean half-planes which are separated by a straight line $l = \mathbb{E}^+ \cap \mathbb{E}^-$ in $\mathbb{E}$.

Definition 1.1 (Upper/Lower Rainbow Block). *Let* $\mathbb{V} = \{v_1, \dots, v_{2n}\} \neq \emptyset$ *be a set of vertices which are ordered on the straight line* $l = \mathbb{E}^+ \cap \mathbb{E}^-$. *We call the resulting upper arch configuration an* upper rainbow block *and denote it by* $\mathscr{B}^+$, *if*

i) (Rainbow Nesting/Rule) *in the upper half-plane* $\mathbb{E}^+$ *each vertex* $v_k \in \mathbb{V}$ *is joined by a half-circle, an* arc, *with* $v_{|\mathbb{V}|-k+1}$, *i.e.,*

$$v_k \textit{ is adjacent to } v_{|\mathbb{V}|-k+1} \textit{ in } \mathbb{E}^+,$$

ii) *all arcs in the upper arch configuration do not intersect each other.*

DOI 10.1515/9783110533026-002

Fig. 1.1: Upper rainbow block $\mathcal{B}^+$ with $|V^{\mathcal{B}^+}| = 6$ and $|A^{\mathcal{B}^+}| = 3$ on the left and lower rainbow block $\mathcal{B}^-$ with $|V^{\mathcal{B}^-}| = 10$ and $|A^{\mathcal{B}^-}| = 5$ on the right.

We denote its non-empty set of vertices by $V^{\mathcal{B}^+}$ *and its non-empty set of arcs by* $A^{\mathcal{B}^+}$*. For the lower rainbow arch configuration* $\mathcal{B}^-$ *replace "* $+$ *" by "* $-$ *" and "upper" by "lower" in the definition of upper rainbow block.*

In each rainbow block $\mathcal{B} = (V^{\mathcal{B}}, A^{\mathcal{B}})$ the following equality holds:

$$|V^{\mathcal{B}}| = 2|A^{\mathcal{B}}|. \tag{1.1}$$

For an illustration of Definition 1.1 and the above equality see Figure 1.1.

Definition 1.2 (Disjoint Upper/Lower Blocks). *Two upper blocks* $\mathcal{B}_k^+$ *and* $\mathcal{B}_l^+$ *are called* disjoint *if their sets of vertices* $V^{\mathcal{B}_k^+}$ *and* $V^{\mathcal{B}_l^+}$ *are disjoint, i.e.,*

$$V^{\mathcal{B}_k^+} \cap V^{\mathcal{B}_l^+} = \emptyset.$$

Analogously, we define disjoint *for two lower blocks.*

We are ready to define *seaweed meanders.*

Definition 1.3 (Seaweed Meander). *Let* $a_1, \dots, a_N$ *represent rainbow upper blocks* $\mathcal{B}_i^+$*,* $i = 1, \dots, N$*, and let* $b_1, \dots, b_M$ *represent rainbow lower blocks*

$$\mathcal{B}_j^-, \; j = 1, \dots, M$$

such that the following two conditions hold:

i)

$$\bigsqcup_{i=1}^{N} V^{\mathcal{B}_i^+} = \bigsqcup_{j=1}^{M} V^{\mathcal{B}_j^-},$$

ii) *the vertices lie on a straight line* l *with an order* $<$ *and* $v \in V^{\mathcal{B}_i^\pm}$*,* $w \in V^{\mathcal{B}_{i+1}^\pm}$ *implies* $v < w$.

By building the upper and lower rainbow blocks in such a way, we obtain an embedded planar graph denoted by

$$\mathcal{S}(a_1, \dots, a_N \mid b_1, \dots, b_M) = (V^{\mathcal{S}}, A^{\mathcal{S}})$$

which we call a seaweed meander.

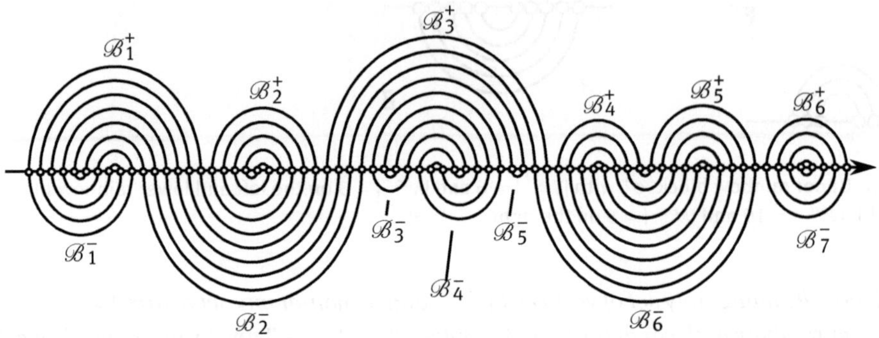

Fig. 1.2: Seaweed meander $\mathcal{S}(8, 5, 10, 4, 5, 4 \mid 5, 10, 2, 4, 1, 10, 4)$ with six upper blocks and seven lower blocks.

Note that a *seaweed meander* consists of finitely many mutually disjoint closed (differentiable) Jordan curves in the plane. The number of Jordan curves – i.e., the number of connected components – of a seaweed meander

$$\mathcal{S}(a_1, \dots, a_r \mid b_1, \dots, b_k)$$

depends on the number of arcs in the upper and lower blocks, i.e., on

$$a_1, \dots, a_r \text{ and } b_1, \dots, b_k.$$

A seaweed meander $\mathcal{S}(a_1, \dots a_r \mid b_1, \dots, b_k)$ decomposes into the disjoint union of two subgraphs:

$$\mathcal{S}(a_1, \dots, a_K \mid b_1, \dots, b_M) = \mathcal{S}(a_1, \dots, a_K \mid b_1, \dots, b_M) \sqcup \mathcal{S}(a_{K+1}, \dots, a_r \mid b_{M+1}, \dots, b_k), \quad (1.2)$$

whenever there exists a $K \in \{1, \dots, r\}$ and a $M \in \{1, \dots, k\}$ such that

$$\sum_{i=1}^{K} a_i = \sum_{j=1}^{M} b_j.$$

In the above example in Figure 1.2 this condition is satisfied.
Namely, it holds that

$$\sum_{i=1}^{5} |\mathscr{B}_i^+| = \sum_{j=1}^{6} |\mathscr{B}_j^-|.$$

Therefore,

$$\mathcal{S}(8, 5, 10, 4, 5, 4 \mid 5, 10, 2, 4, 1, 10, 4) = \mathcal{S}(8, 5, 10, 4, 5 \mid 5, 10, 2, 4, 1, 10) \sqcup \mathcal{S}(4 \mid 4).$$

We observe and prove the following general fact:

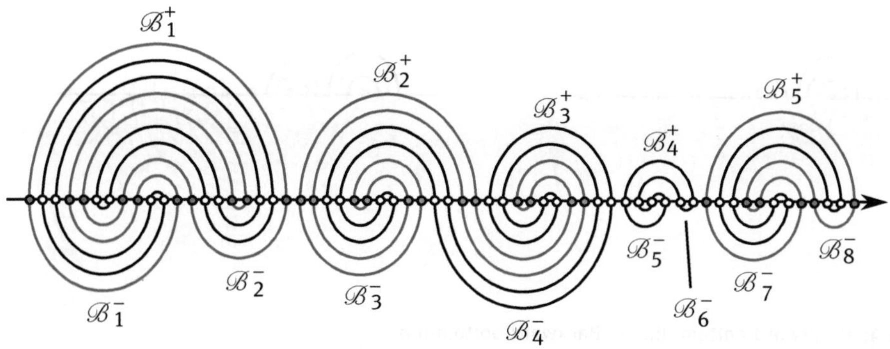

Fig. 1.3: Seaweed meander $\mathcal{S}(10, 7, 5, 3, 6 \mid 6, 4, 5, 7, 2, 1, 4, 2)$.

Lemma 1.1. *Let $\mathcal{S}(a_1, \ldots, a_r \mid b_1, \ldots, b_k)$ be a seaweed meander such that there exist $K_1, \ldots, K_n = r$ and $M_1, \ldots, M_n = k$ such that*

$$\sum_{i=1}^{K_1} a_i = \sum_{j=1}^{M_1} b_j, \quad \sum_{i=K_1+1}^{K_2} a_i = \sum_{j=M_1+1}^{M_2} b_j, \quad \ldots, \quad \sum_{i=K_{n-1}}^{K_n} a_i = \sum_{j=M_{n-1}}^{M_n} b_j.$$

Then, a seaweed meander $\mathcal{S}(a_1, \ldots, a_r \mid b_1, \ldots, b_k)$ decomposes into the disjoint union of n closed subgraphs. In symbols,

$$\begin{aligned}\mathcal{S}(a_1, \ldots, a_r \mid b_1, \ldots, b_k) = \mathcal{S}(a_1, \ldots, a_{K_1} \mid b_1, \ldots, b_{M_1}) \sqcup \\ \mathcal{S}(a_{K_1+1}, \ldots, a_{K_2} \mid b_{M_1+1}, \ldots, b_{M_2}) \sqcup \cdots \sqcup \\ \mathcal{S}(a_{K_{n-1}}, \ldots, a_{K_n} \mid b_{M_{n-1}}, \ldots, b_{M_n}).\end{aligned}$$

Proof. By our assumption, each upper block $\mathscr{B}_{K_1}^+$ has no vertices adjacent to vertices of lower block $\mathscr{B}_{M_1+1}^-$. The same is true for the blocks $\mathscr{B}_{K_2}^+$ and $\mathscr{B}_{M_2+1}^-, \ldots, \mathscr{B}_{K_{n-2}}^+$ and $\mathscr{B}_{M_{n-1}}^-$. Since all upper/lower blocks are a priori disjoint and rainbow without intersecting arcs, we obtain the disjoint union of seaweed meanders as was claimed in the lemma. □

For illustrating the above Lemma 1.1, let us consider the seaweed meander

$$\mathcal{S}(10, 7, 5, 3, 6 \mid 6, 4, 5, 7, 2, 1, 4, 2).$$

By Lemma 1.1, we have

$$\mathcal{S}(10, 7, 5, 3, 6 \mid 6, 4, 5, 7, 2, 1, 4, 2) = \mathcal{S}(10 \mid 6, 4) \sqcup \mathcal{S}(7, 5 \mid 5, 7) \sqcup \mathcal{S}(3 \mid 2, 1) \sqcup \mathcal{S}(6 \mid 4, 2),$$

which is shown in Figure 1.3.

In the following, we introduce a *flip* for a seaweed meander, which we will call a *forward-/backward-bottom-flip* performed on the lower blocks $\mathscr{B}_1^-, \ldots, \mathscr{B}_k^-$ of

$$\mathcal{S}(a_1, \ldots, a_r \mid b_1, \ldots, b_k).$$

The *forward-/backward-bottom-flip* (f-/b-b-f) is also mentioned in [10] and [6].

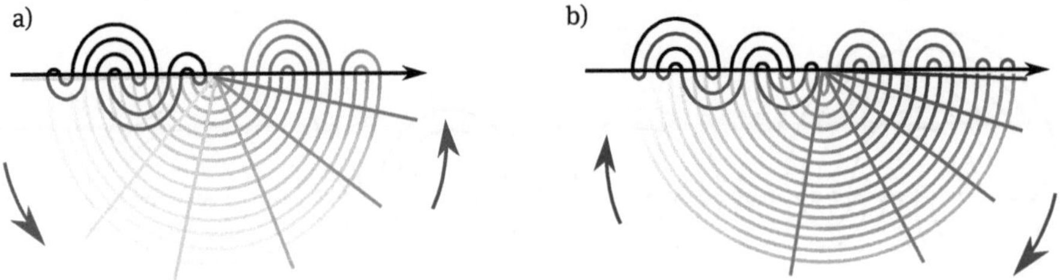

Fig. 1.4: a): Forward-bottom-flip. b): Backward-bottom-flip.

Construction 1 (F-/B-B-F). *Let $\mathcal{S}(a_1, \dots, a_r \mid b_1, \dots, b_k)$ be a seaweed meander.*

1) *Forwards, we pull the lower blocks $\mathscr{B}_1^-, \dots, \mathscr{B}_k^-$ into the upper half-plane $\mathbb{E}^+$ by prolonging the arcs, as indicated in Figure 1.4. After the flip, the number of connected components remains unchanged and the seaweed meander reads as*

$$\mathcal{S}\left(a_1, \dots, a_r, b_k, \dots, b_1 \;\middle|\; \sum_{i=1}^{r} a_i + \sum_{j=1}^{k} b_j\right).$$

In [10] these meanders were called cleaved.

2) *Backwards, the flip is not always as easy as forwards since we may start with a non-cleaved meander*

$$\mathcal{S}\left(a_1, \dots, a_r \;\middle|\; \sum_{i=1}^{r} a_i\right).$$

However, it is possible to choose $K \in \{2, \dots, r-1\}$ such that

$$\sum_{j=1}^{K} a_j \text{ approximately equals to } \sum_{l=K+1}^{r} a_l.$$

Let

$$a_R := \left| \sum_{j=1}^{K} a_j - \sum_{l=K+1}^{r} a_l \right|.$$

We end up with three cases:

a) *If $\sum_{j=1}^{K} a_j > \sum_{l=K+1}^{r} a_l$, then $\mathcal{S}(a_1, \dots, a_K \mid a_r, \dots a_{K+1}, a_R)$.*

b) *If $\sum_{j=1}^{K} a_j < \sum_{l=K+1}^{r} a_l$, then $\mathcal{S}(a_1, \dots, a_K, a_R \mid a_r, \dots a_{K+1})$.*

c) *If $\sum_{j=1}^{K} a_j = \sum_{l=K+1}^{r} a_l$, then $\mathcal{S}(a_1, \dots, a_K \mid a_r, \dots a_{K+1})$.*

By Construction 1 we may introduce seaweed meanders in the simpler notation

$$\mathcal{S}(a_1, \dots, a_N)$$

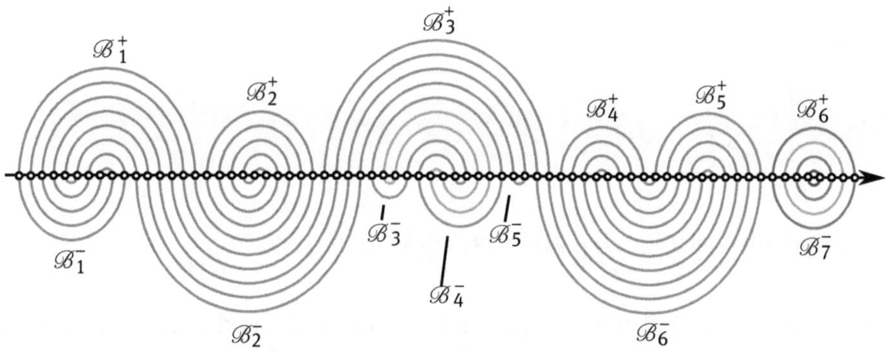

Fig. 1.5: Seaweed meander with colored connected components.

omitting the sum after the "|"-symbol. For simplicity, we denote a seaweed meander by $\mathcal{S}(a_1, \dots, a_r)$. It is given by a composition

$$(a_1, \dots, a_N), \quad a_i \in \mathbb{N}_{>0},$$

where as before $a_i := |\mathrm{A}^{\mathscr{B}_i^+}|$ is the number of arcs in the ith-upper block.

Let us consider the seaweed meander

$$\mathcal{S}(8, 5, 10, 4, 5, 4 \mid 5, 10, 2, 4, 1, 10, 4)$$

in Figure 1.2. We may ask for its number of connected components. The answer can be deduced from the number of colored Jordan curves in Figure 1.5. After the *forward-bottom-flip* we get

$$\mathcal{S}(8, 5, 10, 4, 5, 4, 4, 10, 1, 4, 2, 10, 5),$$

i.e, a seaweed meander with $N = 13$ upper blocks. Let $\mathrm{Z}(a_1, \dots, a_N)$ denote the number of connected components of a seaweed meander $\mathcal{S}(a_1, \dots, a_N)$, where Z stands for *Zusammenhangskomponente German* for connected component.

For $N \in \{1, 2, 3\}$, the following formulae for the number of connected components of $\mathcal{S}(a_1, \dots, a_N)$ are proven, cf. [10]:

$$N = 1: \quad \mathrm{Z}(a_1) = \gcd(a_1); \tag{1.3}$$

$$N = 2: \quad \mathrm{Z}(a_1, a_2) = \gcd(a_1, a_2); \tag{1.4}$$

$$N = 3: \quad \mathrm{Z}(a_1, a_2, a_3) = \gcd(a_1 + a_2, a_2 + a_3). \tag{1.5}$$

However, for $N \geq 4$ any simple formula of the form analogous to (1.5), i.e.,

$$\mathrm{Z}(a_1, \dots, a_N) = \gcd(f_1(a_1, \dots, a_N), f_2(a_1, \dots, a_N))$$

with $f_1, f_2 \in \mathbb{Z}[x]$, remained unknown. Indeed, in the present work we show the following theorem in Section 1.2.2:

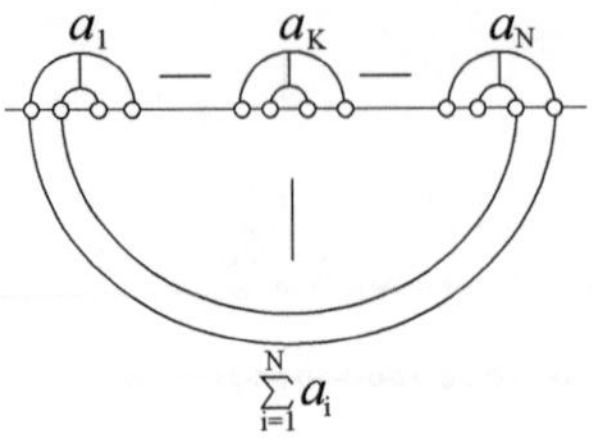

Fig. 1.6: A seaweed meander $\mathcal{S}(a_1, \ldots, a_K, \ldots, a_N)$, $K, N \in \mathbb{N}_{>0}$.

Theorem 1 (Obstruction). Let $a_1, \ldots, a_N \in \mathbb{N}_{>0}$ with $N \geq 4$ and $\mathcal{S}(a_1, \ldots, a_N)$ be a seaweed meander. Then, there do not exist homogeneous polynomials $f_1, f_2 \in \mathbb{Z}[x]$ of arbitrary degree with integer coefficients such that the number of connected components of every seaweed meander $\mathcal{S}(a_1, \ldots, a_N)$ is given by

$$\gcd(f_1(\overline{a}), f_2(\overline{a})) \text{ with } \overline{a} := (a_1, \ldots, a_N).$$

In other words, to every choice of homogeneous polynomials f_1, f_2 we find a counterexample.

In the following, we give the rule for connecting the vertices of a seaweed meander with only one lower block. Let $N \in \mathbb{N}_{>0}$ and $a_1, \ldots, a_N \in \mathbb{N}_{>0}$. We line up $2\sum_{i=1}^{N} a_i =: 2y$ vertices on the x-axis in the Euclidean plane. The vertices are labeled by $v_{a_1}, \ldots, v_{2y}$. Connect two vertices by a *half-circle* (an *arc*) in the following way:
above the straight line:

$$\text{for } j \in \{1, \ldots, N\},\ k \in \{1, \ldots, a_j\} : v_{2\sum_{i=1}^{j-1} a_i + k} \quad \text{with} \quad v_{2\sum_{i=1}^{j} a_i + 1 - k}$$

below the straight line:

$$\text{for } k \in \{1, \ldots, N\} : v_k \quad \text{with} \quad v_{2\sum_{i=1}^{N} a_i + 1 - k}\ .$$

By following the above rules, we obtain N *upper blocks* $\mathcal{B}_1^+, \ldots, \mathcal{B}_N^+$ and one *lower block* $\mathcal{B}^-$. Each ith-upper block consists of a_i-many and the bottom block $\mathcal{B}^-$ of y-many simply nested arcs. By this rule, we get an embedded planar graph $\mathcal{S}(a_1, \ldots, a_N)$ called a *seaweed meander*, see Figure 1.6. We call the N-tuple $(a_1, \ldots, a_N)$ a *composition* of a seaweed meander $\mathcal{S}(a_1, \ldots, a_N)$.

1.1.1 Collapse Procedure for Seaweed Meanders

In this section, we introduce the *collapse* of seaweed meanders. The collapse translates seaweed meanders into what we call *collapsed seaweed meanders*. Some connectivity properties of seaweed meanders are more apparent after this translation. In particular, we derive a necessary property of connected seaweed meanders which will be used for the proof of Theorem 1. To begin with, the number of vertices in a seaweed meander $\mathcal{S}(a_1, \ldots, a_N)$ equals to $2\sum_{i=1}^{N} a_i$, which is even. This is a consequence of Formula 1.1. By using Formula 1.1, we may glue each vertex v_i with its right neighbor v_{i+1} by starting with the first vertex from the left, i.e.,

$$v_i \quad \text{with} \quad v_{i+1}, \qquad i \in \left\{1,\ 3,\ 5,\ \ldots,\ 2\left(\sum_{i=1}^{N} a_i\right) - 1\right\}.$$

By this procedure two neighboring arcs of each block in the upper half-plane $\mathbb{E}^+$, respectively, in the lower half-plane $\mathbb{E}^-$, collapse to one arc. There is one exceptional case in which the number of arcs in an upper block is odd, i.e.,

$$|\mathrm{A}^{\mathscr{B}_k^+}| = a_k = 2l + 1.$$

If the number of arcs of a rainbow block is odd, then only one arc remains in the middle and therefore collapses to one single vertex. We call the middle vertices of each odd a_i *semi-isolated*, i.e., there is no arc on top emanating from it. In the sequel, we denote collapsed seaweed meanders by $\mathcal{S}^{coll}(a_1, \ldots, a_N)$ and call them *collapsed seaweed meanders.* Note, that in $\mathcal{S}^{coll}(a_1, \ldots, a_N)$ each a_i corresponds to the number of vertices of ith-upper block. For an illustration see Figure 1.7. In $\mathcal{S}^{coll}(a_1, \ldots, a_N)$ the number of arcs in each ith-upper block reduces to $\frac{a_i}{2}$ for even a_i and to $\left\lfloor \frac{a_i}{2} \right\rfloor$ for odd a_i after the collapse. After the collapse, the number of vertices of each ith-upper block reduces as well, namely to $|\mathrm{A}^{\mathscr{B}_i^+}|$. Therefore, we end up with an embedded planar graph given by

$$\mathcal{S}^{coll} = \left(\mathrm{V}^{\mathcal{S}^{coll}}, \mathrm{A}^{\mathcal{S}^{coll}}\right) \text{with } |\mathrm{V}^{\mathcal{S}^{coll}}| = |\mathrm{A}^{\mathcal{S}}|$$

and similarly for $i, j \in \{1, \ldots N\}$:

$$|\mathrm{A}^{\mathcal{S}^{coll}}| = \overbrace{\sum_{a_i\ odd} \left\lfloor \frac{a_i}{2} \right\rfloor + \sum_{a_j\ even} \frac{a_j}{2}}^{\text{total number of arcs in } \mathbb{E}^+} + \underbrace{\left\lfloor \frac{\sum_{i=1}^{N} a_i}{2} \right\rfloor}_{\text{total number of arcs in } \mathbb{E}^-}. \tag{1.6}$$

Based on the Jordan curve theorem we introduce the following construction.

Construction 2 (Collapse of Seaweed Meanders). *Let $\mathcal{S}(a_1, \ldots, a_N)$ be a seaweed meander. Each connected component of a seaweed meander is a closed Jordan curve in the plane. Hence, we may color the interior of each connected component. By this we obtain thick components which are by zooming out collapsed seaweed meanders denoted by $\mathcal{S}^{coll}(a_1, \ldots, a_N)$ constructed as follows:*

1) *On the straight line in $\mathbb{E}$, draw the vertices in each corresponding collapsed upper block $\overline{\mathscr{B}}_1^+, \ldots, \overline{\mathscr{B}}_N^+$, i.e., for each upper block $\overline{\mathscr{B}}_i^+$ draw a_i-many vertices. For the collapsed lower block $\overline{\mathscr{B}}^-$ the vertices are automatically drawn.*
2) *In $\mathbb{E}^+$ connect in each collapsed upper block $\overline{\mathscr{B}}_i^+$ the vertices by an arc following the rainbow nesting/rule, see Definition 1.1, i). Connect the vertices in the same way in $\mathbb{E}^-$.*

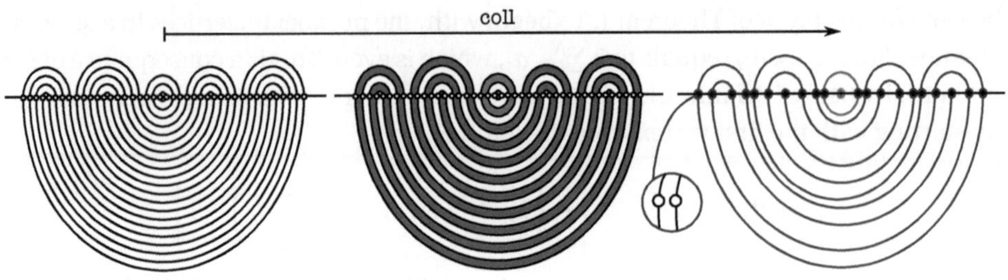

Fig. 1.7: Left: seaweed meander $\mathcal{S}(4, 5, 3, 4, 5)$. Middle: the coloring of the interior regions of connected components in $\mathcal{S}(4, 5, 3, 4, 5)$. Right: collapsed seaweed meander $\mathcal{S}^{coll}(4, 5, 3, 4, 5)$. The number of connected components of $\mathcal{S}(4, 5, 3, 4, 5)$ is four: one component with blue interior and three components encapsulating in the middle. In $\mathcal{S}^{coll}(4, 5, 3, 4, 5)$ there are two paths, namely one in blue and one isolated point in green surrounded by the red cycle in the middle. The colors are left out in the abstract meander graph.

In short, the collapse of seaweed meanders is given by the following map:

$$\begin{aligned} \texttt{coll} : \mathscr{S} &\to \mathscr{S}^{coll} \\ \mathcal{S}(a_1, \ldots, a_N) &\mapsto \texttt{coll}(\,\mathcal{S}(a_1, \ldots, a_N)) =: \mathcal{S}^{coll}(a_1, \ldots, a_N), \end{aligned} \tag{1.7}$$

where $\mathscr{S}$ denotes the set of seaweed meanders and $\mathscr{S}^{coll}$ the set of the collapsed seaweed meanders constructed following Construction 2. Basically, the map `coll` switches the meaning of the numbers a_i, $i = 1, \ldots, a_N$. Before the collapse the entries count the arcs, whereas after the map was applied they count the nodes, which is again by Construction 2.

Since to each collapsed seaweed meander there is exactly one seaweed meander, the map `coll` is injective. Since every seaweed meander is collapsible, the map `coll` is surjective. Therefore, the map `coll` is invertible, meaning that there is the inverse map

$$\begin{aligned} \texttt{coll}^{-1} : \mathscr{S}^{coll} &\to \mathscr{S} \\ \mathcal{S}^{coll}(a_1, \ldots, a_N) &\mapsto \texttt{coll}^{-1}(\,\mathcal{S}^{coll}(a_1, \ldots, a_N)) = \mathcal{S}(a_1, \ldots, a_N). \end{aligned} \tag{1.8}$$

By the one-to-one correspondence established by the map `coll`, the sets $\mathscr{S}^{coll}$ and $\mathscr{S}$ are of the same cardinality, i.e., $|\mathscr{S}^{coll}| = |\mathscr{S}|$. Since it is possible to generalize the above map `coll` for meanders later on, the enumerative questions considered for example in [8] are sufficient to be answered for either collapsed or non-collapsed meanders.

The advantage of the collapse of seaweed meanders lies in the following: After the collapse we obtain two types of connected components, namely cycles and paths including *isolated* points, which are semi-isolated on top and bottom. Isolated points are paths of length zero.

The isolated points are formed by the following condition on $a_i's$:

$$\sum_{i=1}^{K} a_i = \sum_{i=K+2}^{N} a_i, \qquad K \in \{1, \ldots, N-2\}. \tag{1.9}$$

We illustrate the above Construction 2 and Theorem 1.1 by applying the collapse map `coll` to the seaweed meander $\mathcal{S}(4, 5, 3, 4, 5)$ in Figure 1.7 on p. 14.

Theorem 1.1. *Let* $\mathrm{Z}(a_1, \ldots, a_N)$ *denote the number of connected components of a seaweed meander* $\mathcal{S}(a_1, \ldots, a_N)$, $\mathrm{C}(a_1, \ldots, a_N)$ *the number of cycles and* $\mathrm{P}(a_1, \ldots, a_N)$ *the number of paths of collapsed seaweed meander* $\mathcal{S}^{coll}(a_1, \ldots, a_N)$*. Then, the number of connected components of a seaweed meander* $\mathcal{S}(a_1, \ldots, a_N)$ *equals to twice the number of cycles plus the number of paths in* $\mathcal{S}^{coll}(a_1, \ldots, a_N)$*. In symbols,*

$$\mathrm{Z}(a_1, \ldots, a_N) := 2\,\mathrm{C}(a_1, \ldots, a_N) + \mathrm{P}(a_1, \ldots, a_N). \tag{1.10}$$

Proof. Since connected components of a seaweed meander $\mathcal{S}(a_1, \ldots, a_N)$ are closed Jordan curves, each connected component has exactly one interior region, which is by the Jordan curve theorem. Therefore, we may color each interior region in $\mathcal{S}(a_1, \ldots, a_N)$ and hence produce thick connected components of three types, namely paths, cycles and isolated points. The boundaries of cycles are two connected components of $\mathcal{S}(a_1, \ldots, a_N)$ and therefore must be counted twice, whereas the boundaries of paths and isolated points are formed by only one connected component and hence must be counted once. □

By the above Theorem 1.1, it suffices to count the number of cycles and paths of the collapsed meander in order to determine the number of connected components of a seaweed meander. This perspective allows us to obtain results in the next section.

We note that it is not only possible to introduce the collapse for meanders, i.e., meanders containing non-rainbow blocks, but also to introduce the so called *multiple collapse*. We first establish the multiple collapse for seaweed meanders $\mathcal{S}(a_1, \ldots, a_N)$ with $a_i, i = 1, \ldots, N$, such that a_i are divisible by $\lambda > 1$ and then generalize it for meanders in Section 2.1.

Construction 3 (Multiple Collapse). *Let* $\mathcal{S}(a_1, \ldots, a_N)$ *be a seaweed meander with* $a_i = \lambda b_i,\ i = 1, \ldots, N,\ \lambda \in \mathbb{N}_{>1}$*. Identify* λ*-many arcs by starting with the outermost one. After identifying we obtain a seaweed meander* $\mathcal{S}(b_1, \ldots, b_N)$*. By collapsing* $\mathcal{S}(b_1, \ldots, b_N)$ *we obtain the multiple collapse of* $\mathcal{S}(a_1, \ldots, a_N)$*.*

We introduce the multiple collapse by the following map, for some examples consult Figure 1.8:

$$\begin{aligned} \mathtt{Coll} : \mathscr{S} &\to \mathscr{S}^{coll} \\ \mathcal{S}(\lambda b_1, \ldots, \lambda b_N) &\mapsto \mathtt{Coll}\,\big(\,\mathcal{S}(\lambda b_1, \ldots, \lambda b_N)\,\big) = \mathcal{S}^{coll}(b_1, \ldots, b_N). \end{aligned} \tag{1.11}$$

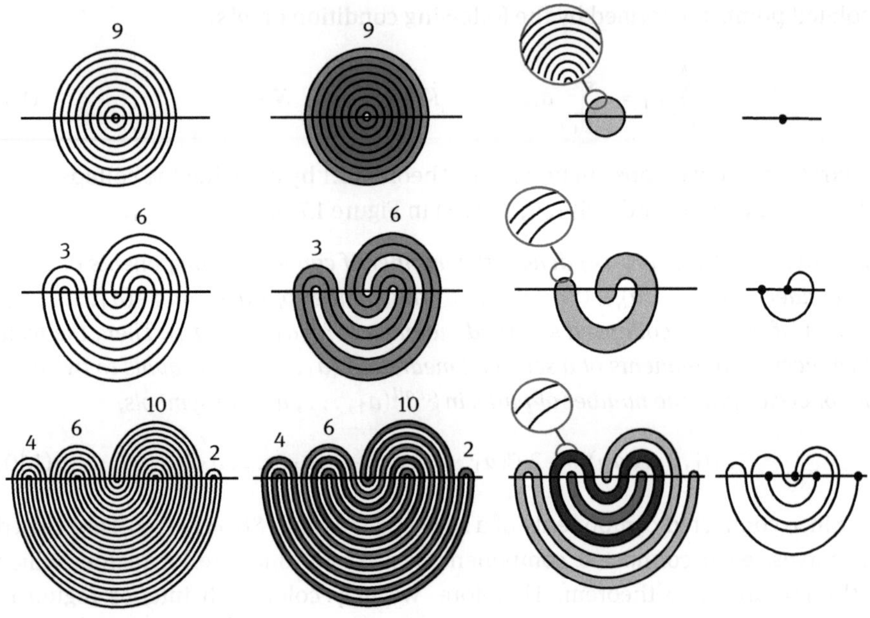

Fig. 1.8: Multiple Collapse.

Since we had glued instead of two neighboring arcs, as in case of collapse, λ-many arcs with each other in the multiple collapse, the formula for the number of connected components obtained in Theorem 1.1 must scale by λ. This is equivalent to the linear-scaling-property proven in Lemma 1.5.

1.1.2 Collapse: Results on Connectivity

In this section we collect some results used in the proof of Theorem 1. Most of the results are based on the idea of the *collapse* of seaweed meanders.

Lemma 1.2 (Number of Paths). *Let $\mathcal{S}(a_1, \ldots, a_N)$ be a seaweed meander and κ the number of odd a_i's in $(a_1, \ldots, a_N)$. Then the number of paths $\mathrm{P}(a_1, \ldots, a_N)$ in the corresponding collapsed seaweed meander $\mathcal{S}^{coll}(a_1, \ldots, a_N)$ equals to $\left\lceil \frac{\kappa}{2} \right\rceil$.*

Proof. The proof splits into two cases.

Case 1: The number of odd a_i's is even, i.e.,

$$\kappa = 2m \text{ with } m \in \mathbb{N}_0.$$

Then each path has its endvertices in the middle points of two upper blocks with an odd number of arcs, as the other upper blocks with an even number of vertices have no semi-isolated points from above and all arcs in $\mathcal{S}^{coll}(a_1, \ldots, a_N)$ are rainbow. Therefore, the number of paths is exactly m.

Case 2: The number of odd a_i's is odd, i.e.,

$$\kappa = 2m + 1 \text{ with } m \in \mathbb{N}_0.$$

Then, as for the previous case, m paths have their endvertices in $2m$ middle points such that one middle point remains. Either it is an isolated point (if condition (1.9) holds) or it is an endvertex of a path with an additional endvertex

$$v_{\left\lceil \frac{\sum_{i=1}^{N} a_i}{2} \right\rceil}.$$

Therefore, the number of paths is $m + 1 = \left\lceil \frac{\kappa}{2} \right\rceil$. □

From the above Lemma 1.2, we deduce following corollaries:

Corollary 1.1 (Parity). *Let $\mathcal{S}(a_1, \ldots, a_N)$ be a connected seaweed meander. Then the number of odd a_i's is at least one but at most two.*

Proof. The number of connected components grows by Theorem 1.10 and Lemma 1.2, i.e.,

$$\mathrm{Z}(a_1, \ldots, a_N) = 2\,\mathrm{C}(a_1, \ldots, a_N) + \mathrm{P}(a_1, \ldots, a_N) = 2\mathrm{C}(a_1, \ldots, a_N) + \left\lceil \frac{\kappa}{2} \right\rceil,$$

where κ denotes the number of odd a_i's in $\mathcal{S}^{coll}(a_1, \ldots, a_N)$. Hence, compositions $(a_1, \ldots, a_N)$ of connected meanders must satisfy $\mathrm{C}(a_1, \ldots, a_N) = 0$, i.e.,

$$\mathcal{S}^{coll}(a_1, \ldots, a_N)$$

is cycle-free and

$$\mathrm{P}(a_1, \ldots, a_N) = \left\lceil \frac{\kappa}{2} \right\rceil = 1,$$

i.e., κ is one or two. □

Remarks 1.1.

i) *Note, that if we consider a connected seaweed meander*

$$\mathcal{S}\left(a_1, \ldots, a_N \,\middle|\, \sum_{i=1}^{N} a_i\right)$$

then there must be exactly two odd numbers in the set $\{a_1, \ldots, a_N, \sum_{i=1}^{N} a_i\}$. Similarly, if a connected seaweed meander is given by

$$\mathcal{S}(a_1, \ldots, a_N \mid b_1, \ldots, b_M)$$

then there must be exactly two odd numbers in $\{a_1, \ldots, a_N, b_1, \ldots, b_M\}$.

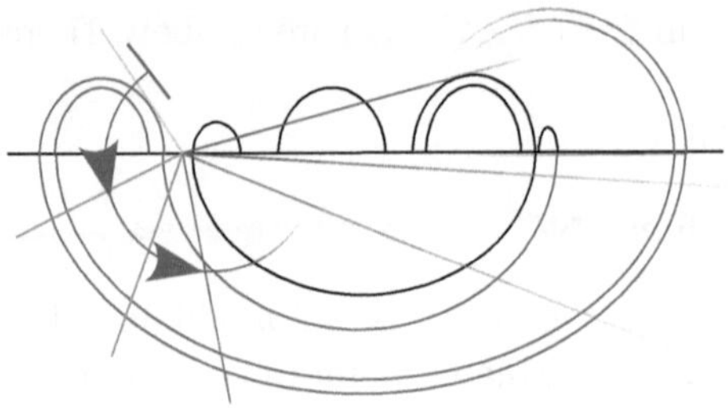

Fig. 1.9: In red: forward-bottom-flip for $\mathcal{S}(a+1, a, c, 2a+2)$ performed on the first upper block with $2|A^{\mathcal{B}_1^+}| = |A^{\mathcal{B}_4^+}|$. In black: obtained seaweed meander $\mathcal{S}(a, c, a+1)$.

ii) *Corollary 1.1 will be generalized for meanders with the aid of the collapse in the next chapter.*

iii) *The above corollary was formulated for the index of seaweed Lie algebras and proven by D. I. Panyushev in [24, 4.7 Corollary]. In comparison to Panyushev's proof of 4.7 Corollary we used different arguments and formulation, since we proved it before we were aware of the relation to the index of seaweed Lie algebras to be discussed in the next chapter.*

Corollary 1.2 (Odd Number of Connected Components). *Let a seaweed meander*

$$\mathcal{S}(a_1, \ldots, a_N)$$

contain one or two odd a_i's. Then, the number of connected components in $\mathcal{S}(a_1, \ldots, a_N)$ is odd.

Proof. By Lemma 1.2 the number of paths, in symbols $\mathrm{P}(a_1, \ldots, a_N)$, in a seaweed meander $\mathcal{S}^{coll}(a_1, \ldots, a_N)$ with one or two odd a_i's equals to one. Therefore, the number of connected components in $\mathcal{S}^{coll}(a_1, \ldots, a_N)$ is

$$\mathrm{Z}(a_1, \ldots, a_N) = 2\,\mathrm{C}(a_1, \ldots, a_N) + 1,$$

which is odd. □

Lemma 1.3. *Let $N = 4$ and $\mathcal{S}(a+1, a, c, 2a+2)$ with $a, c \in \mathbb{N}_{>0}$. Then, $\mathcal{S}(a_1, \ldots, a_N)$ is connected, i.e.,*

$$\mathrm{Z}(a+1, a, c, 2a+2) = 1.$$

Proof. Since $2a_1 = 2(a+1) = a_4$, we can perform the forward-bottom-flip on the first upper block with $|A^{\mathcal{B}_1^+}| = a + 1$ arcs as shown in Figure 1.9 without changing the number of connected components. We reduce

$$(a+1, a, c, 2a+2) \text{ to } (a, c, a+1).$$

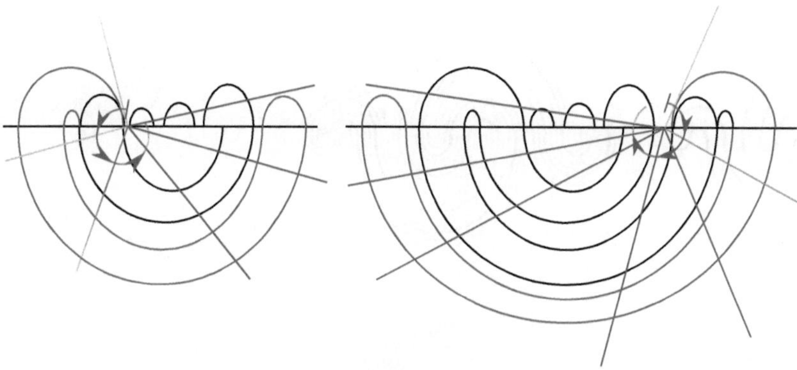

Fig. 1.10: Illustration of Lemma 1.4; left: forward-flip; right: backward-flip.

By using Formula (1.5), it follows that

$$Z(a, c, a + 1) = \gcd(a + c, a + c + 1) = 1. \qquad \square$$

The following lemma serves for the generalization of Lemma 1.3.

Lemma 1.4. *Let* $4 < N \in \mathbb{N}_{>0}$ *and* $\alpha := a + 1$, $a \in \mathbb{N}_{>0}$. *Then the following holds:*

$$\text{If } N \text{ is even,} \quad \mathcal{S}^{coll}(\underbrace{\alpha, \underbrace{2\alpha, \ldots, 2\alpha}_{\frac{N-2}{2}}, a, c, \underbrace{2\alpha, \ldots, 2\alpha}_{\frac{N-4}{2}}}_{N}) \text{ is connected.}$$

$$\text{If } N \text{ is odd,} \quad \mathcal{S}^{coll}(\underbrace{\underbrace{2\alpha, \ldots, 2\alpha}_{\lceil\frac{N-4}{2}\rceil}, a, c, \underbrace{2\alpha, \ldots, 2\alpha}_{\lceil\frac{N-4}{2}\rceil}, \alpha}_{N}) \text{ is connected.}$$

Proof. Apply the forward-/backward-flip on the upper block with α arcs in order to extend $(\alpha, a, c, 2\alpha)$. $\square$

Remark. *For all* $N \in \mathbb{N}_{>0}$ *and arbitrary* $a, c \in \mathbb{N}_{>0}$, $\alpha := a + 1$ *we obtain*

$$1 = Z(\underbrace{\alpha, \underbrace{2\alpha, \ldots, 2\alpha}_{\frac{N-2}{2}}, a, c, \underbrace{2\alpha, \ldots, 2\alpha}_{\frac{N-4}{2}}}_{N \text{ even}}) = Z(\underbrace{\underbrace{2\alpha, \ldots, 2\alpha}_{\lceil\frac{N-4}{2}\rceil}, a, c, \underbrace{2\alpha, \ldots, 2\alpha}_{\lceil\frac{N-4}{2}\rceil}, \alpha}_{N \text{ odd}}) = 1.$$

Note that depending on $a, c \in \mathbb{N}_{>0}$ *all compositions contain at least one but at most two odd numbers. This argument will be used in the generalization of the proof of Theorem 1.2 in Steps 1, 2.*

The following result plays an important role in the proof of Theorem 1 and is valid for both seaweed meanders and meanders.

Fig. 1.11: Illustration of the linear-scaling-property: a) seaweed meanders $\mathcal{S}(1, 4, 2 \mid 2, 4, 1)$ and $\mathcal{S}(4, 16, 8 \mid 8, 16, 4)$; b) meanders $\mathcal{M}(2, 2(2, 2), 1 \mid 1, 2(2, 2), 2)$ and $\mathcal{M}(6, 6(6, 6), 3 \mid 3, 6(6, 6), 6)$.

Lemma 1.5 (Linear-Scaling-Property). *Let*

$$\mathcal{S}(\lambda a_1, \ldots, \lambda a_N)$$

be a seaweed meander and $\lambda \in \mathbb{N}$. Then, the number of connected components scales linearly, i.e.,

$$\mathrm{Z}(\lambda a_1, \ldots, \lambda a_N) = \lambda \mathrm{Z}(a_1, \ldots, a_N).$$

Proof. Assume that $\lambda > 1$, otherwise there is nothing to prove. Let $M \in \mathbb{N}_{>0}$ be the number of connected components of a seaweed meander. Each connected component of $\mathcal{S}(a_1, \ldots, a_N)$ is a cycle

$$C_i \text{ for all } i = 1, \ldots, M$$

with non-intersecting arcs. Therefore, the set of vertices of $\mathcal{S}$ – in symbols $\mathrm{V}^{\mathcal{S}}$ – is partitioned into M disjoint subsets

$$\mathrm{V}_1, \ldots, \mathrm{V}_M, \text{ i.e., } \mathrm{V}^{\mathcal{S}} = \bigsqcup_{i=1}^{M} \mathrm{V}_i.$$

By multiplying the number of arcs in each block by λ, we obtain instead of one arc λ-many simply nested arcs emanating in $\mathbb{E}^+$ and $\mathbb{E}^-$ from each original vertex. Therefore, the length of each cycle with multiple arcs

$$\overline{C}_i \text{ for all } i = 1, \dots, M,$$

which is obtained from C_i, $i = 1, \dots, M$, by copying it λ-many times is the same as for C_i, $i = 1, \dots, M$, i.e., the lengths of cycles are preserved after the scaling. In other words, in order to obtain a scaled seaweed meander we replace each original vertex by λ-many vertices where the arcs in $\mathbb{E}^+$ and $\mathbb{E}^-$ touch each other on the straight line. Hence, the total number of cycles of $\mathcal{S}(\lambda a_1, \dots, \lambda a_N)$ is λM, as was claimed. □

We illustrate the linear-scaling-property by the following Figure 1.11. Note the four copies on the right of the original cycle on the left in a) and three copies on the right of the original cycle on the left in b).

Remarks 1.2.

i) *The linear-scaling-property also holds for* meanders *to be introduced in the next section.*

ii) *Similar to the Corollary 1.1, the above lemma was formulated and differently proven in [24, 4.4 Corollary, ii)], again for the index of seaweed Lie algebras. We will explain in the chapter on seaweed Lie algebras why this correspondencies occurred.*

1.2 Proof of the Obstruction-Theorem

As we have indicated before, we shall prove the theorem which states that there is no possibility to generalize the formula for the number of connected components of seaweed meanders with at least four upper blocks in terms of the greatest common divisor and homogeneous polynomials of arbitrary degree. Our result was formulated as Conjecture 19 in [6] and posed as a question in [10]. We start with the proof for seaweed meanders with four upper blocks and then give its generalization to arbitrary number of upper blocks.

1.2.1 Obstruction-Theorem for Seaweed Meanders: Four Upper Blocks

Theorem 1.2 (Obstruction, $N = 4$). *Let $a_1, \dots, a_4 \in \mathbb{N}_{>0}$, $N = 4$ and $\mathcal{S}(a_1, \dots, a_4)$ be a seaweed meander. Then, there do not exist homogeneous polynomials $f_1, f_2 \in \mathbb{Z}[x]$ of arbitrary degree with integer coefficients such that the number of connected components of every seaweed meander $\mathcal{S}(a_1, \dots, a_4)$ is given by*

$$\gcd(f_1(\overline{a}), f_2(\overline{a})) \text{ with } \overline{a} := (a_1, \dots, a_4).$$

In other words, to every choice of homogeneous polynomials f_1, f_2 we find a counter-example.

Proof.

Step 1. We show that one of the degrees must be one, i.e., $\deg(f_1) = 1$ or $\deg(f_2) = 1$. Let f_1, f_2 be homogeneous polynomials of degree

$$d_i := \deg(f_i) \in \mathbb{N}, i \in \{1, 2\},$$

i.e.,

$$\forall \lambda \in \mathbb{Z} : f_i(\lambda x_1, \dots, \lambda x_4) = \lambda^{d_i}(x_1, \dots, x_4).$$

Choose $\lambda \in \mathbb{Z}$ coprime to $f_1(\overline{a}), f_2(\overline{a})$. Then,

$$\gcd(f_1(\lambda\overline{a}), f_2(\lambda\overline{a})) = \lambda^{\min(d_1,d_2)}\gcd(f_1(\overline{a}), f_2(\overline{a})). \tag{1.12}$$

Since by the linear-scaling-property in Lemma 1.5 the number of connected components of a seaweed meander $\mathcal{S}(a_1, \dots, a_N)$ scales linearly, i.e., the following equality holds:

$$\mathrm{Z}(\lambda\overline{a}) = \lambda\mathrm{Z}(\overline{a}), \tag{1.13}$$

from (1.12) and (1.13), it follows that f_1 or f_2 has to be of degree one.

Step 2. We show that both degrees are one, i.e., $\deg(f_1) = 1 = \deg(f_2)$. Without loss of generality assume $d_2 > 1$ and $d_1 = 1$. From Lemma 1.4

$$\mathcal{S}^{coll}(a + 1, a, c, 2a + 2)$$

is connected and hence 1.1 $\mathcal{S}(a + 1, a, c, 2a + 2)$ is connected. Therefore, there exist

$$a, c \in \mathbb{N}_{>0} \text{ such that } f_1(a + 1, a, c, 2a + 2) \neq 1.$$

Set $1 \neq \lambda = f_1(a + 1, a, c, 2a + 2)$. Then consider,

$$\begin{aligned}\lambda &= \mathrm{Z}(\lambda(a + 1, a, c, 2a + 2)) \\ &= \gcd(\lambda f_1(a + 1, a, c, 2a + 2), \lambda^{d_2} f_2(a + 1, a, c, 2a + 2)) \\ &= \lambda^2 \gcd(1, \lambda^{d_2-2} f_2(a + 1, a, c, 2a + 2)) \\ &= \lambda^2,\end{aligned}$$

which is a contradiction. Therefore, $d_2 = 1$. *Step 3. We derive conditions on the coefficients of* f_1, f_2. From *Step 1* and *Step 2* we know that f_i, $i = 1, 2$, are homogeneous polynomials of degree one. Define,

$$f_1 := \sum_{i=1}^{4} \alpha_i x_i \text{ and } f_2 := \sum_{i=1}^{4} \beta_i x_i, \ \alpha_i, \ \beta_i \in \mathbb{Z}.$$

Further, let

$$S_1 := \{2k_1 + 1, 2k_2 + 1, 2k_3, 2k_4\}$$

be a set with two odd and two even positive (non-zero) natural numbers and

$$S_2 := \{2k_1 + 1, 2k_2, 2k_3, 2k_4\}$$

Fig. 1.12: 2 × 4-Board.

be a set with one odd and three even positive (non-zero) natural numbers. Then, from Lemma 1.2 the number of connected components of seaweed meanders

$$\mathcal{S}(a_1, a_2, a_3, a_4) \text{ such that } a_i \in S_1 \text{ or } a_i \in S_2,$$

must be odd , i.e., for all $s, j \in \{1, \dots, 4\}$, $s \neq j$, $L, M \in \mathbb{N}$

$$\begin{aligned} 2L + 1 &= \mathcal{Z}(a_1, \dots, a_4) \\ &= \gcd\Big(2\sum_{i=1}^{4} \alpha_i k_i + \alpha_s + \alpha_j, 2\sum_{i=1}^{4} \beta_i k_i + \beta_s + \beta_j\Big), \quad a_i \in S_1 \qquad (1.14) \\ 2M + 1 &= \mathcal{Z}(a_1, \dots, a_4) \\ &= \gcd\Big(2\sum_{i=1}^{4} \alpha_i k_i + \alpha_j, 2\sum_{i=1}^{4} \beta_i k_i + \beta_j\Big), \quad a_i \in S_2. \qquad (1.15) \end{aligned}$$

must be satisfied. Therefore, we obtain the following conditions on α_i, β_i after plugging $a_i \in S_i$ into the polynomials f_i, $i \in \{1, 2\}$: For all $j, s \in \{1, \dots, 4\}$, $s \neq j$:

$$\alpha_j + \alpha_s \equiv 0 \ (mod\ 2) \text{ and } \beta_j + \beta_s \equiv 1 \ (mod\ 2) \qquad (1.16)$$

$$\alpha_j + \alpha_s \equiv 1 \ (mod\ 2) \text{ and } \beta_j + \beta_s \equiv 0 \ (mod\ 2) \qquad (1.17)$$

$$\alpha_j + \alpha_s \equiv 1 \ (mod\ 2) \text{ and } \beta_j + \beta_s \equiv 1 \ (mod\ 2) \qquad (1.18)$$

$$\alpha_j \equiv 1 \ (mod\ 2) \text{ and } \beta_j \equiv 0 \ (mod\ 2) \qquad (1.19)$$

$$\alpha_j \equiv 0 \ (mod\ 2) \text{ and } \beta_j \equiv 1 \ (mod\ 2) \qquad (1.20)$$

$$\alpha_j \equiv 1 \ (mod\ 2) \text{ and } \beta_j \equiv 1 \ (mod\ 2). \qquad (1.21)$$

Step 4. We argue that such conditions can not be fulfilled for $N = 4$. In order to show that the above conditions on α_i, β_i can't be fulfilled for $N = 4$, consider a (2×4)-board with 2 rows and 4 columns, see Figure 1.12

The first row corresponds to α_i and the second to β_i, $i \in \{1, \dots, 4\}$. Let us represent even α, β by a black square "■" and odd by a white square "□". Due to the conditions (1.19)–(1.21) we are allowed to put vertically the following (2×1)-stones on the (2×4)-board:

"■□", "□■" and "□□".

By the pigeonhole principle one of the stones appears twice on the board, contradicting conditions (1.16)–(1.18). □

The proof of the generalization of the above theorem – for seaweed meanders with arbitrary number (greater than four) of upper blocks – is same to the above proof in its structure.

1.2.2 Obstruction-Theorem for Seaweed Meanders with at Least Four Upper Blocks

Theorem 1.3 (Obstruction, $N \geq 4$). *Let $a_1, \dots, a_N \in \mathbb{N}_{>0}$, $N \geq 4$ and $\mathcal{S}(a_1, \dots, a_N)$ be a seaweed meander. Then, there do not exist homogeneous polynomials $f_1, f_2 \in \mathbb{Z}[x]$ of arbitrary degree with integer coefficients such that the number of connected components of every seaweed meander $\mathcal{S}(a_1, \dots, a_N)$ is given by*

$$\gcd(f_1(\overline{a}), f_2(\overline{a})) \text{ with } \overline{a} := (a_1, \dots, a_N).$$

In other words, to every choice of homogeneous polynomials f_1, f_2 we find a counter-example.

Proof. *Step 1. We show that one of the degrees must be one, i.e.,*

$$\deg(f_1) = 1 \text{ or } \deg(f_2) = 1$$

Let f_1, f_2 be homogeneous polynomials of degree $d_i := \deg(f_i) \in \mathbb{N}$, $i \in \{1, 2\}$, i.e.,

$$\forall \lambda \in \mathbb{Z} : f_i(\lambda x_1, \dots, \lambda x_N) = \lambda^{d_i}(x_1, \dots, x_N).$$

Choose $\lambda \in \mathbb{Z}$ coprime to $f_1(\overline{a}), f_2(\overline{a})$. Then,

$$\gcd(f_1(\lambda\overline{a}), f_2(\lambda\overline{a})) = \lambda^{\min(d_1,d_2)}\gcd(f_1(\overline{a}), f_2(\overline{a})). \tag{1.22}$$

Since the number of connected components of a seaweed meander scales linearly, i.e., the following equality holds:

$$\mathrm{Z}(\lambda\overline{a}) = \lambda\mathrm{Z}(\overline{a}), \tag{1.23}$$

from (1.22) and (1.23), it follows that f_1 or f_2 has to be of degree one.

Step 2. We show that both degrees are one, i.e., $deg(f_1) = 1 = deg(f_2)$. Without loss of generality assume $d_2 > 1$ and $d_1 = 1$. Define

$$A_{\text{even}} := (a + 1, 2a + 2, \dots, 2a + 2, a, c, 2a + 2, \dots, 2a + 2).$$

From Lemma 1.4 we know that for arbitrary $a, c \in \mathbb{N}_{>0}$ and an even N the following holds: $\mathcal{S}^{coll}(A_{\text{even}})$ is connected and hence by Theorem 1.1 $\mathcal{S}(A_{\text{even}})$ is connected. Therefore, there exist

$$a, c \in \mathbb{N}_{>0} \text{ such that } f_1(A_{\text{even}}) \neq 1.$$

Set $1 \neq \lambda = f_1(A_{\text{even}})$. Then, the linear-scaling-property of Z in (1.23) and homogeneity of f_1, f_2 yield

$$\lambda = \mathrm{Z}(\lambda A_{\text{even}}) = \gcd(\lambda f_1(A_{\text{even}}), \lambda^{d_2} f_2(A_{\text{even}})) = \lambda^2 \gcd(1, \lambda^{d_2-2} f_2(A_{\text{even}})) = \lambda^2,$$

which is a contradiction. Analogously, we can argue for an odd $N \geq 4$ with

$$A_{\text{odd}} := (2a+2, \ldots, 2a+2, a, c, 2a+2, \ldots, 2a+2, a+1)$$

and obtain a contradiction. Therefore,

$$d_2 = 1 \quad \text{for } N \geq 4.$$

Step 3. We derive conditions on the coefficients of f_1, f_2. From *Step 1, 2* we know that $f_i,\ i = 1, 2$, are homogeneous polynomials of degree one. Define,

$$f_1 := \sum_{i=1}^{N} \alpha_i x_i \text{ and } f_2 := \sum_{i=1}^{N} \beta_i x_i,\ \ \alpha_i,\ \beta_i \in \mathbb{Z}.$$

Further, let

$$S_1 := \{2k_1 + 1, 2k_2 + 1, 2k_3, \ldots, 2k_N\}$$

be a set with two odd and $N-2$ even numbers and

$$S_2 = \{2k_1 + 1, 2k_2, 2k_3, \ldots, 2k_N\}$$

be a set with one odd and $N-1$ even numbers, where $k_i \in \mathbb{N}_{>0}$. Then, from Lemma 1.2 the number of connected components of seaweed meanders formed by compositions $(a_1, \ldots, a_N)$, such that

$$a_i \in S_1 \text{ or } a_i \in S_2,\ \ i = 1, \ldots, N$$

must be odd, i.e., for all $s, j \in \{1, \ldots, N\}, s \neq j, L, M \in \mathbb{N}$ holds:

$$\begin{aligned} 2L + 1 &= Z(a_1, \ldots, a_4) \\ &= \gcd\Big(2\sum_{i=1}^{N} \alpha_i k_i + \alpha_s + \alpha_j, 2\sum_{i=1}^{N} \beta_i k_i + \beta_s + \beta_j\Big),\ \ a_i \in S_1 \qquad (1.24) \\ 2M + 1 &= Z(a_1, \ldots, a_4) \\ &= \gcd\Big(2\sum_{i=1}^{N} \alpha_i k_i + \alpha_j, 2\sum_{i=1}^{N} \beta_i k_i + \beta_j\Big),\ \ a_i \in S_2. \qquad (1.25) \end{aligned}$$

Therefore, we obtain the following conditions on α_i, β_i after plugging $a_i \in S_i$ into the polynomials $f_i,\ i \in \{1, 2\}$: For all $j, s \in \{1, \ldots, N\}, s \neq j$:

$$\alpha_j + \alpha_s \equiv 0 \ (mod\ 2) \text{ and } \beta_j + \beta_s \equiv 1 \ (mod\ 2) \qquad (1.26)$$

$$\alpha_j + \alpha_s \equiv 1 \ (mod\ 2) \text{ and } \beta_j + \beta_s \equiv 0 \ (mod\ 2) \qquad (1.27)$$

$$\alpha_j + \alpha_s \equiv 1 \ (mod\ 2) \text{ and } \beta_j + \beta_s \equiv 1 \ (mod\ 2) \qquad (1.28)$$

$$\alpha_j \equiv 1 \ (mod\ 2) \text{ and } \beta_j \equiv 0 \ (mod\ 2) \qquad (1.29)$$

$$\alpha_j \equiv 0 \ (mod\ 2) \text{ and } \beta_j \equiv 1 \ (mod\ 2) \qquad (1.30)$$

$$\alpha_j \equiv 1 \ (mod\ 2) \text{ and } \beta_j \equiv 1 \ (mod\ 2). \qquad (1.31)$$

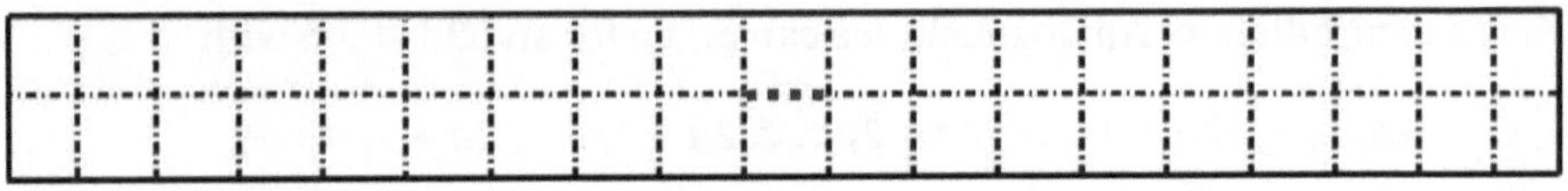

Fig. 1.13: $2 \times N$-Board.

Step 4. We argue that such conditions can not be fulfilled for $N \geq 4$. In order to show that the above conditions on α_i, β_i can't be fulfilled for $N \geq 4$, consider a $(2 \times N)$-board with 2 rows and N columns, see Figure 1.13.

The first row corresponds to α_i and the second to β_i, $i \in \{1, \dots, N\}$. Let us represent even α, β by a black square "■" and odd by a white square "□". Due to the conditions (1.29)–(1.31) we are allowed to put vertically the following (2×1)-stones on the $(2 \times N)$-board:

"■□" "□■" and "□□".

By the pigeonhole principle one of the stones appears twice on the board, contradicting conditions (1.26)–(1.28). □

Remark. *The above proof may be generalized in order to exclude formulae of the form* $\gcd(f_1, \dots, f_k), f_j \in \mathbb{Z}[x], j = 1, \dots, k$ *for seaweed meanders with at least* 2^k *upper rainbow blocks. By considering a* $(N \times k)$*-board and the pigeonhole principle in Step 4, we again deduce a contradiction.*

2 Meanders

We introduce *meanders* proceeding similarly to the previous section. We will also collapse *meanders* in order to study connectivity. The reason for considering the generalization of seaweed meanders is that *Morse meanders* may be viewed as collapsed meanders with additional requirements on the lower and upper *general blocks*. Morse meanders will be introduced later on in Chapter 3 and play a key-role in the description of global Sturm attractors, see next three chapters.

Definition 2.1 (Non-Rainbow Upper/Lower Blocks). *We call an upper block*

$$\mathscr{B}^+ = (\mathrm{V}^{\mathscr{B}^+}, \mathrm{A}^{\mathscr{B}^+}) \quad \textit{with} \quad \mathrm{V}^{\mathscr{B}^+} \neq \emptyset, \ \mathrm{A}^{\mathscr{B}^+} \neq \emptyset,$$

a non-rainbow upper block, *if v_1 is adjacent to $v_{2|\mathrm{A}^{\mathscr{B}^+}|}$, each vertex is adjacent to one other vertex (in other words each vertex has degree one), all arcs (semi-cycles) are non-oriented and pairwise non-intersecting and* non-rainbow, *i.e., there exists at least one*

$$v_K \in V^{\mathscr{B}^+}, \ K \in \{2, \dots, 2|\mathrm{A}^{\mathscr{B}^+}| - 1\}$$

such that it is not adjacent to $v_{2|\mathrm{A}^{\mathscr{B}^+}|-K+1}$. Similarly, we define a non-rainbow *lower block*

$$\mathscr{B}^- = (\mathrm{V}^{\mathscr{B}^-}, \mathrm{A}^{\mathscr{B}^-}) \textit{ in } \mathbb{E}^-.$$

As for the rainbow blocks, we require the set of arcs and vertices of non-rainbow blocks to be non-empty.

Definition 2.2 (General Upper/Lower Block). *We call an upper block*

$$\mathscr{B}^+ = (\mathrm{V}^{\mathscr{B}^+}, \mathrm{A}^{\mathscr{B}^+})$$

general *or simply upper* block, *if it is either rainbow or non-rainbow. Similarly, we define a (general) lower* block

$$\mathscr{B}^- = (\mathrm{V}^{\mathscr{B}^-}, \mathrm{A}^{\mathscr{B}^-}) \textit{ in } \mathbb{E}^-.$$

Note, that for non-rainbow blocks the following lemma is valid:

Lemma 2.1 (Decomposition of Non-Rainbow Blocks). *Let*

$$\mathscr{B}^+ = (\mathrm{V}^{\mathscr{B}^+}, \mathrm{A}^{\mathscr{B}^+})$$

be a non-rainbow upper block. Then, the vertex set $\mathrm{V}^{\mathscr{B}^+}$ is partitioned into vertex sets

$$\mathrm{V}^{\mathscr{B}_1^+}, \dots, \mathrm{V}^{\mathscr{B}_n^+}$$

for which the vertices fulfill:

$$v_i \textit{ is adjacent to } v_{\alpha_k - i+1}, \textit{ where } \alpha_k = |\mathrm{A}^{\mathscr{B}_k^+}|, k = 1, \dots, n$$

and $\mathscr{B}^+$ is a rainbow block. In other words, any non-rainbow block decomposes into the union of rainbow blocks.

DOI 10.1515/9783110533026-003

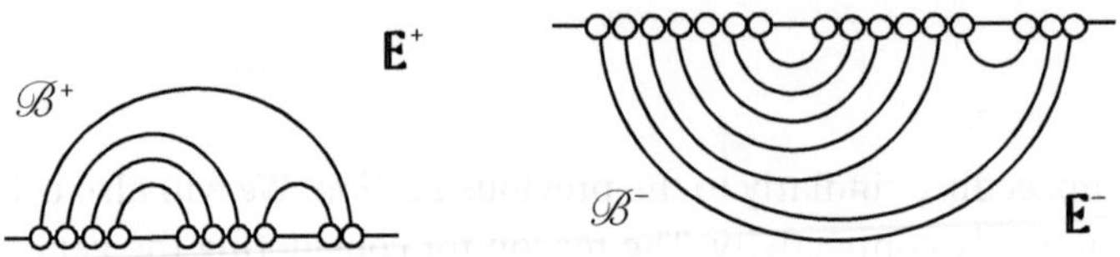

Fig. 2.1: Left: non-rainbow upper block; right: non-r-ainbow lower block.

Proof. By the property of a *non-rainbow* block there is at least one v_k in the vertex set such that it is not adjacent to $v_{2|A^{\mathscr{B}^+}|-k+1}$. Since all vertices are pairwise adjacent in a block, there is a vertex

$$v_{2|A^{\mathscr{B}^+}|-m+1} \quad \text{with} \quad m \neq k.$$

Then, the $(2|A| - m - 1)$-many vertices between

$$v_k \text{ and } v_{2|A^{\mathscr{B}^+}|-m+1}$$

must be connected by an arc beneath the arc joining

$$v_k \text{ with } v_{2|A^{\mathscr{B}^+}|-m+1},$$

since otherwise we would produce an intersection of arcs. Therefore, either all vertices between them are joined by rainbow arcs, i.e., the block beneath the arc connecting

$$v_k \text{ with } v_{2|A^{\mathscr{B}^+}|-m+1}$$

is rainbow or there is a vertex

$$v_L \quad \text{with} \quad L \in \{k+1, \ldots, 2|A^{\mathscr{B}^+}| - m\}$$

such that it is not adjacent to $v_{2|A^{\mathscr{B}^+}|-L+1}$. Again, all arcs joining the vertices in between must lie beneath the arc joining the two vertices. By finiteness of the vertex set we will finally reach the last pair of vertices. Since the arcs lie beneath the arc we have started with and depending on how many of them are rainbow we get the desired claim. □

We illustrate the above Definition 2.2 by Figure 2.1 and introduce the notation for non-rainbow blocks, since it is not sufficient to represent them by only one number of arcs as for the rainbow blocks. Our aim will be to represent the blocks by ordered weighted rooted trees.

Definition 2.3 (i^{th}-Level of a Non-Rainbow Block). *Let*

$$i = 0, \ldots, L \geq 1$$

and $\mathscr{B}$ be a general block. The ith-level of a general block consists of the outermost rainbow blocks lying under the rainbow blocks contained in the $(i-1)^{st}$-level.

Notation 1 (Non-Rainbow Blocks – Bracketing). *Let $\mathcal{B}_l$ be a non-rainbow block with L-many levels.*

1) *We enumerate the rainbow blocks in the corresponding levels*

$$0, \ldots, L$$

by starting with 0th-level containing the first rainbow block denoted by $\mathcal{B}_l^1$ and write down the maximal number of arcs contained in it, i.e.,

$$|\mathrm{A}^{\mathcal{B}_l^1}| = a^0.$$

Since there is only one outermost rainbow block in the 0th-level we are done.

2) *In the first level, we enumerate the rainbow blocks by*

$$\mathcal{B}_1^1, \ldots, \mathcal{B}_{k(1)}^1$$

and write

$$a_1^1, \ldots, a_{k(1)}^1$$

for the corresponding number of arcs. Analogously, we proceed for the remaining levels

$$2, \ldots, L.$$

3) *We represent the non-rainbow nesting of a block by writing down the maximal numbers of arcs contained in the outermost rainbow blocks in each level. By opening brackets we indicate the beginning of a new level. Explicitly, the bracket expression of an arbitrary non-rainbow block is given as follows:*

$$\begin{aligned}
&\textit{Level } 0: \quad a^0\\
&\textit{Level } 1: \quad a^0(a_1^1 \ldots, a_{k(1)}^1)\\
&\textit{Level } 2: \quad a^0(a_1^1(a_1^{2,1}, \ldots, a_{k(2,1)}^{2,1}), \ldots, a_{k(1)}^1(a_1^{2,k(1)}, \ldots, a_{k(2,k(1))}^{2,k(1)})\\
&\quad\vdots\\
&\textit{Level } L: \quad a^0\Big(a_1^1(a_1^{2,1}(\ldots(a_{\cdot}^{L,\cdot}, \ldots, a_{\cdot}^{L,\cdot})\ldots), \ldots,\\
&\qquad\qquad a_{k(2,1)}^{2,1}(\ldots(a_{\cdot}^{L,\cdot}, \ldots, a_{\cdot}^{L,\cdot}), \ldots)), \ldots,\\
&\qquad\quad a_{k(1)}^1(a_1^{2,k(1)}(\ldots(a_{\cdot}^{L,\cdot}, \ldots, a_{\cdot}^{L,\cdot})\ldots), \ldots,\\
&\qquad\qquad a_{k(2,k(1))}^{2,k(1)}(\ldots(a_{\cdot}^{L,\cdot}, \ldots, a_{\cdot}^{L,\cdot})\ldots))\Big),
\end{aligned}$$

where "$\cdot$" in the upper respectively lower indices of a stands for omitting the indices.

For each bracket expression A representing a block with K-many levels we can associate the *canonical tree* T_A. The canonical tree

$$\mathrm{T}_A = (\mathrm{V}^{\mathrm{T}_A},\ \mathrm{E}^{\mathrm{T}_A},\ r,\ \omega : \mathrm{V}^{\mathrm{T}_A} \to \mathbb{N}_{>0})$$

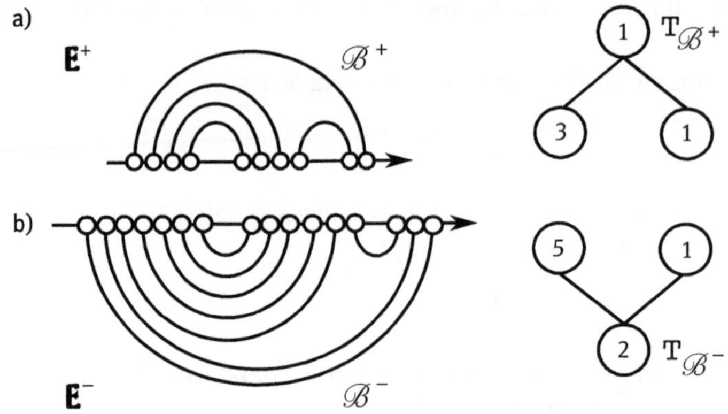

Fig. 2.2: a) Upper and b) lower block with the associated canonical trees.

is a rooted weighted tree with the root r and the weight function ω on the vertex set $\mathrm{V}^{\mathrm{T}_A}$. With respect to each level $k = 0, \ldots, K-1$ of T_A we define the (linear) *tree order* in the following way:

Definition 2.4 (Tree Order). *Let A and T_A be as above. A tree order $<$ is a linear order on each set of vertices*

$$\mathrm{V}_k^{\mathrm{T}_A} := \{v \in \mathrm{V}^{\mathrm{T}_A} \mid \mathtt{dist}(v, r) = k\},$$

of the kth-level of T_A such that if

$$x, y \in \mathrm{V}_k^{\mathrm{T}_A} \text{ and } x', y' \in \mathrm{V}_{k+1}^{\mathrm{T}_A} \text{ with } xx', yy' \in \mathrm{E}^{\mathrm{T}_A} \text{ then } x < y \text{ implies } x' < y'.$$

We say that $<$ defines a tree order on T_A.

Notation 2 (Ordered Rooted Weighted Trees for Blocks). *Let $\mathcal{B}$ be a block given by the bracket expression in the above notation with L-many levels. We assign to $\mathcal{B}$ an ordered rooted tree $\mathrm{T}_{\mathcal{B}}$ by translating canonically the levels of $\mathcal{B}$ into the levels of a tree, namely by interpreting level-wise the numbers in front of the brackets as weighted parents and the numbers inside of the brackets as its weighted children. Parents and children in $\mathrm{T}_{\mathcal{B}}$ are written from left to right by respecting the order of the bracketing expression of $\mathcal{B}$.*

Definition 2.5 (Canonical Tree). *We call ordered rooted weighted trees $\mathrm{T}_{\mathcal{B}}$ as introduced in the Notation 2 for non-rainbow blocks $\mathcal{B}$* canonical trees.

Remark. *Up to two levels in the Notation 1, a canonical tree $\mathrm{T}_{\mathcal{B}}$ grows as follows: The 0th-level of the tree $\mathrm{T}_{\mathcal{B}}$ consists of the root with weight a^0 which is the maximal number of outermost rainbow arcs contained in the 0th-level of $\mathcal{B}$. The 1st-level of the tree $\mathrm{T}_{\mathcal{B}}$ consists of the children*

$$v_1^1, \ldots, v_{k(1)}^1 \text{ with weights } a_1^1, \ldots, a_{k(1)}^1$$

corresponding to the root r with weight a^0. *Let* v_j^1 *be the vertex with weight*

$$a_j^1,\ j = 1, \ldots, k(1)\ in\ \mathrm{T}_{\mathscr{B}}.$$

It is important to note, that in the tree the vertices are ordered as follows

$$v_1^1 < v_2^1 < \ldots < v_{k(1)}^1,$$

which is by the tree order. For the remaining levels we proceed recursively. The 2nd *- level of the tree* $\mathrm{T}_{\mathscr{B}}$ *consists of the children*

$$v_1^1, \ldots, v_{k(1)}^1 \text{ with weights } a_1^{2,i}, \ldots, a_{k(2,i)}^{2,i}$$

corresponding to the parents with weights a_i^1, $i = 1, \ldots, k(1)$ *of the previous level.*

It is easy to prove by induction, that bracket expressions and canonical trees corresponding to blocks are in one-to-one correspondence.

For example, let us consider Figure 2.2. We have two levels for both $\mathscr{B}^+$ and $\mathscr{B}^-$. In the 0th-level, there is one arc and in 1st-level there are two rainbow blocks: one rainbow block with three arcs and the other with one arc. Therefore, we write the arcs in the upper block in Figure 2.2 as $1(3, 1)$. In similar way, $2(5, 1)$ stands for the nesting of arcs in the lower block $\mathscr{B}^-$ (in the same figure). The associated trees $\mathrm{T}_{\mathscr{B}^+}$ and $\mathrm{T}_{\mathscr{B}^-}$ are shown in Figure 2.2.

Definition 2.6 (Meander). *Let* $A_1, \ldots, A_N$ *represent general upper blocks*

$$\mathscr{B}_i^+, i = 1, \ldots, N$$

and let $B_1, \ldots, B_M$ *represent general lower blocks* $\mathscr{B}_j^-$, $j = 1, \ldots, M$ *such that the following two conditions hold:*

i)

$$\bigsqcup_{i=1}^{N} V^{\mathscr{B}_i^+} = \bigsqcup_{j=1}^{M} V^{\mathscr{B}_j^-}$$

ii) *the vertices lie on a straight line* l *with an order*[1] $\prec$ *and satisfy*

$$v \in \mathrm{V}^{\mathscr{B}_i^\pm}, w \in \mathrm{V}^{\mathscr{B}_{i+1}^\pm} \text{ implies } v \prec w.$$

By building the upper and lower general blocks in such a way, we obtain an embedded planar graph denoted by

$$\mathcal{M}(A_1, \ldots, A_N \mid B_1, \ldots, B_M) = (\mathrm{V}^{\mathcal{M}}, \mathrm{A}^{\mathcal{M}})$$

which we call a meander.

1 For v_k, $v_l \in \mathrm{V}^{\mathscr{B}_i^\pm}$, $v_k \prec v_l$: v_k precedes v_l on the straight line.

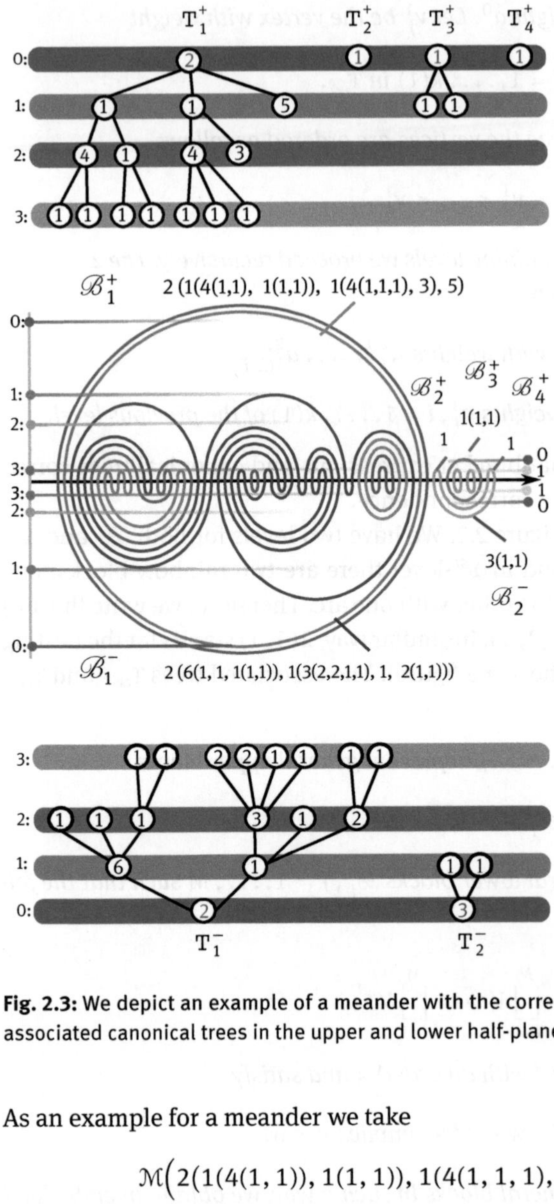

Fig. 2.3: We depict an example of a meander with the corresponding levels in $\mathbb{E}^{+}$ and $\mathbb{E}^{-}$ and its associated canonical trees in the upper and lower half-planes.

As an example for a meander we take

$$\mathcal{M}\Big(2(1(4(1,1)),1(1,1)),1(4(1,1,1),3),5),\ 1,\ 1(1,1),\ 1\ \Big|\\ 2(6(1,1,1(1,1)),1(3(2,2,1,1),1,2(1,1)),\ 3(1,1)\Big)$$

shown in Figure 2.3.

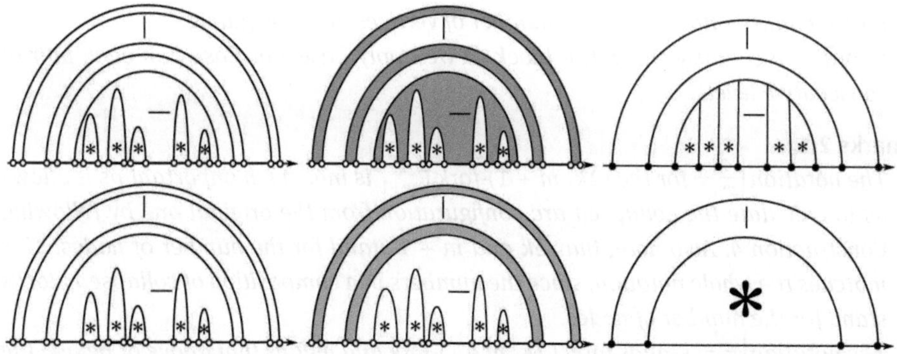

Fig. 2.4: Gluing of the arcs contained in the first two levels of a block is shown in the middle of each row. Top row: The zeroth level contains odd number of arcs. Bottom row: The number of arcs in the zeroth level is even.

2.1 Collapse of (Closed) Meanders

In order to introduce the collapse for meanders we first consider the collapse of an upper/lower block $\mathscr{B}$. As for seaweed meanders we will glue together two neighboring vertices and the corresponding pair of incident arcs.

Construction 4 (Collapse of a Non-Rainbow Block). *Let $\mathscr{B}$ be a non-rainbow block in $\mathbb{E}^+$ or in $\mathbb{E}^-$ and let*

$$a^1(\,a^{2,1}(\dots),\ \dots\ ,a^{2,m}(\dots)\,)$$

represent its nesting (cf. Notation 1). For the collapse of $\mathscr{B}$ we distinguish between two cases:

1) *a^1 is odd, i.e $a^1 = 2k + 1$,*
2) *a^1 is even, i.e $a^1 = 2k$.*

To case 1): *If $a^1 = 2k + 1$, then by gluing together the first and the last k pairs of vertices we glue the first incident k pairs of arcs. In the middle, one arc remains and is glued with the first m arcs of the second level. In such a way we obtain an arc with m bars in the middle as shown in Figure 2.4. We call the obtained collapsed configuration a $(2k, m + 1)$-fork and denote it by F^{2k}_{m+1}. For its representation we write $\left|\frac{2k}{m+1}\right.$. Since the first outermost arcs of the second level were glued to the last arc of the first level, the numbers of arcs in the second level reduce by one, i.e.,*

$$a^{2,1} - 1,\ \dots\ ,a^{2,m} - 1.$$

Note, that zero is allowed and is even.

To case 2): *If $a^1 = 2k$, then we obtain k outermost simply nested arcs after the gluing of 2k arcs of the first level. The numbers*

$$a^{2,1},\ \dots\ ,a^{2,m}$$

do not change and become the number of vertices after the gluing.
In order to collapse the entire block $\mathscr{B}$ we apply Case 1 or Case 2 in each pair of consecutive levels.

Remarks 2.1.

i) *The notation* $\Big|\frac{2k}{m+1}$ *for the* $(2k, m+1)$*-fork* F^{2k}_{m+1} *is inasmuch important as it allows us to calculate the collapsed arc configuration from the original one by following Construction 4. Also note, that* $2k$ *and* $m+1$ *stand for the number of nodes. This matches the whole notation, since the numbers in a composition of collapsed blocks stand for the number of nodes.*

ii) *The notation* $\Big|\frac{2k}{m+1}$ *stands for a* $(2k, m+1)$*-fork and means that above or below*[2] *the middle branched part with* $(m+1)-2$ *bars lie k simply nested arcs (i.e in a rainbow configuration).*

iii) *If there are only zeros appearing in the brackets of* $\Big|\frac{2k}{m+1}$ *we can omit them, i.e.,*

$$\Big|\frac{2k}{m+1}(0,\ldots,0) = \Big|\frac{2k}{m+1},$$

since by $m+1$ *we know that we have* $m-1$ *bars in its middle part with empty spaces in between.*

Example 2.1 (Collapse of Blocks).

a) *Consider* $\mathscr{B}_2$ *with* $1(1,1,1,1,1,1,1,1,1,1,1)$*. After the collapse (compare with Case 1 in Construction 4) we obtain* $\Big|\frac{0}{12}(0,0,0,0,0,0,0,0,0,0,0) = \Big|\frac{0}{12}$ *which stands for the* $(0,12)$*-fork* F^0_{12}*.*

b) *Consider* $\mathscr{B}_1$ *given by*

$$1\big(2(3,2(1,1),4),\ 1(4,1,2(1,1))\big).$$

In order to collapse the complete non-rainbow block $\mathscr{B}_1$ *we apply Case 1 or Case 2 of Construction 4 in each level, i.e.,*

$$
\begin{aligned}
\textit{Level } 1,2: \quad & \Big|\tfrac{0}{3}\big([2-1](\ldots),\ [1-1](\ldots)\big) = \Big|\tfrac{0}{3}\big(1(\ldots),\ 0(\ldots)\big)\\
\textit{Level } 2,3: \quad & \Big|\tfrac{0}{3}\big(\Big|\tfrac{0}{4}(3-1,[2-1](\ldots),4-1),\ 0(4,1,2(\ldots)\big) =\\
& \Big|\tfrac{0}{3}\big(\Big|\tfrac{0}{4}(2,1(\ldots),3),\ 0(4,1,2(\ldots))\big)\\
\textit{Level } 3,4: \quad & \Big|\tfrac{0}{3}\big(\Big|\tfrac{0}{4}(2,\Big|\tfrac{0}{3}(1-1,1-1),3),\ 0(4,1,2(1,1))\big) =\\
& \Big|\tfrac{0}{3}\big(\Big|\tfrac{0}{4}(2,\Big|\tfrac{0}{3}(0,0),3),\ 0(4,1,2(1,1))\big) =\\
& \Big|\tfrac{0}{3}\big(\Big|\tfrac{0}{4}(2,\Big|\tfrac{0}{3},3),\ 4,1,2(1,1)\big).
\end{aligned}
$$

There are two $(0,3)$*-forks* F^0_3 *and one* $(0,4)$*-fork* F^0_4*.*

---

2 This depends on whether the underlying fork is contained in a collapsed upper or lower block in $\mathbb{E}^+$ or in $\mathbb{E}^-$.

a)

b)

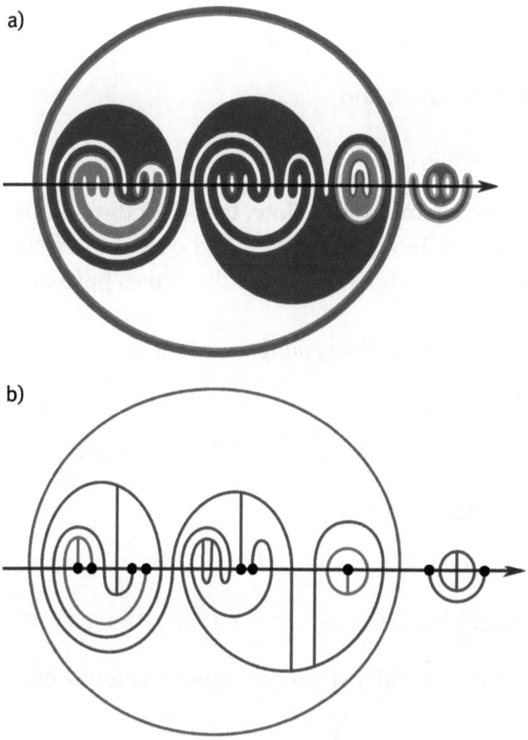

Fig. 2.5: a): Meander from p. 32 with colored regions; b): its collapse.

We are ready to give the following definition:

Definition 2.7 (Collapsed Meander). *Let* $\mathcal{M}(\cdot \mid \cdot)$ *be a meander with* $\mathscr{B}_1^+, \ldots, \mathscr{B}_m^+$ *upper and* $\mathscr{B}_1^-, \ldots, \mathscr{B}_n^-$ *lower blocks. By collapsing each upper and lower block as described in the Construction 4, we obtain a structure*[3] *which we call the* collapsed meander *and denote it by* $\mathcal{M}^{coll}(\cdot \mid \cdot)$, *where "* $\cdot$ *" stands for the collapsed upper blocks before "|" and for the collapsed lower blocks after "|" in the notation of Construction 4.*

In short, we introduce the following map for the collapse of meanders:

$$\begin{aligned} \mathtt{coll} : \mathscr{M} &\to \mathscr{M}^{coll} \\ \mathcal{M}(A_1, \ldots, A_N \mid B_1, \ldots, B_M) &\mapsto \mathcal{M}^{coll}(\overline{A}_1, \ldots, \overline{A}_N \mid \overline{B}_1, \ldots, \overline{B}_M), \end{aligned} \tag{2.1}$$

3 Strictly speaking, we do *not* have a meander anymore due to the existence of forks for the odd case. Due to the existence of forks which include branched arcs, we need to mark the branching points by a vertex in order to be able to view the collapsed meanders as embedded planar graphs. Since we do not need to take the vertices corresponding to the branching points into account in our further investigations, we may ignore them.

where

$$A_i, i = 1, \dots, N \text{ and } B_j, j = 1, \dots, M$$

are bracket expressions each representing a block and

$$\overline{A}_i, i = 1, \dots, N \text{ and } \overline{B}_j, j = 1, \dots, M$$

are bracket expressions in the notation of Construction 4. Note, that after the collapse a bracket expression stay the same except its interpretation[4] if and only if its canonical tree contain only even internal nodes, i.e, all nodes except the leaves have even weights.

Similar to the preceding chapter we define *multiple collapse* as follows:

$$\begin{gathered}\mathtt{Coll} : \mathscr{M} \to \mathscr{M}^{\mathtt{coll}} \\ \mathcal{M}(\beta A_1, \dots, \beta A_N \mid \beta B_1, \dots, \beta B_M) \mapsto \mathcal{M}^{coll}(\overline{A}_1, \dots, \overline{A}_N \mid \overline{B}_1, \dots, \overline{B}_M),\end{gathered} \tag{2.2}$$

where A_i, B_j, $\overline{A}_i$, $\overline{B}_j$ are same as in the definition of the map `coll`.

Example 2.2. *We consider the collapse of the following meander:*

$$\mathcal{M}\Big(2(1(4(1,1)),1(1,1)),1(4(1,1,1),3),5),\ 1,\ 1(1,1),\ 1\ \Big|2(6(1,1,1(1,1)),1(3(2,2,1,1),1,2(1,1)),\ 3(1,1)\Big),$$

from the previous section in Figure 2.3, p. 32. By collapsing each upper and lower block we get:

$$\mathcal{M}^{coll}\left(2\left(\big|^{0}_{3}\left(\big|^{2}_{3},1,1,\big|^{0}_{3}\left(\big|^{2}_{4},2\right),\ 5\right),1,\big|^{0}_{3},1\ \right|\ 2\left(6\left(1,1,\big|^{0}_{3}\right),\big|^{0}_{4}\left(2\left(2,2,1,1\right),0,\big|^{0}_{3}\right)\right),\big|^{2}_{3}\right),$$

which is depicted in Figure 2.5.

As was mentioned before, the only structural difference between the collapse of rainbow and non-rainbow blocks is the existence of forks.

With the aid of Construction 4, we consider all possible cases occurring in the collapse of a meander containing a fork. Up to homotopic deformations we end up with two different cases, shown in Figure 2.6, p. 37.

The existence of forks in the collapse of a meander is exactly the reason why we need to introduce the following:

Definition 2.8 (Chamber). *Let F^{2k}_{m+1} be a $(2k, m+1)$-fork occurring in a collapsed meander. Assume that a pair of distinct vertices*

$$v_i, v_j \in \{v_{k+1}, \dots, v_{k+m}\}$$

of F^{2k}_{m+1} is joined in such a way that a bounded region is produced. We call a bounded region obtained in that way a chamber.

4 In this case, the numbers stand for the number of nodes after the collapse.

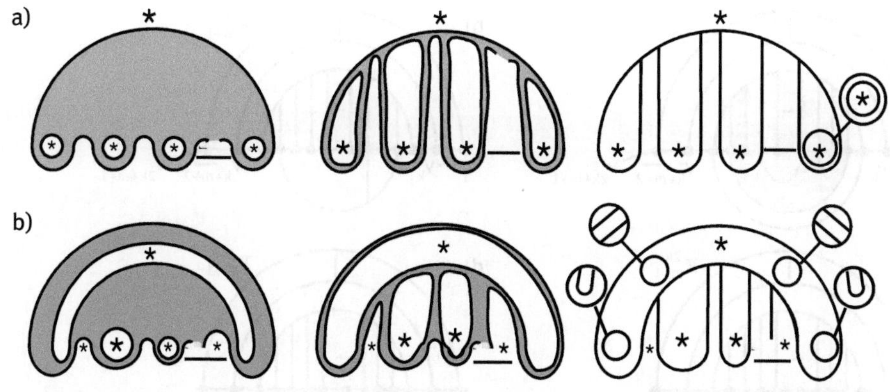

Fig. 2.6: Visualization of two possible cases a) and b) in which chambers emerge up to homotopic deformations. From left to right: coloring of the interior; gluing; schematic drawing.

Example 2.3 (Chambers). *Let F^{2k}_{m+1} be a $(2k, m+1)$-fork in $\mathbb{E}^{+/-}$ and let*

$$v_1, \ldots, v_{2k+m+1}$$

be the corresponding vertices. Assume, that there exists a pair of adjacent vertices

$$v_{k+i} \text{ and } v_{k+j}, \quad i \neq j, \quad i, j \in \{1, \ldots, m+1\},$$

i.e., there is an arc in $\mathbb{E}^{-/+}$ joining two vertices of the branched arc of a $(2k, m+1)$-fork. We obtain a bounded region – a chamber *– see Figure 2.7, a)–d), for its illustration. In Figure 2.7, a) we depict a chamber by joining the nodes*

$$v_{k+3} \text{ and } v_{k+4}$$

in a $(2k, m+1)$-fork. The dashed lines are indicating the other possible closings of a pair of bars and thus forming different chambers. Note, that joining two nodes by a fork is also possible. If a pair of bars of these two forks share a node two chambers are formed, see Figure 2.7, b). In Figure 2.7, c) all vertices corresponding to the bars are not connected to other bars and we have only one chamber. The last picture of Figure 2.7 is probably the most interesting one. Here, we have a chamber colored in gray which is joined to the middle part of a $(2k, m+1)$-fork which contains a light purple chamber in its middle part.

Our goal for the rest of this section consists in establishing the correspondence between the number of connected components of a meander and its collapsed form, similar to Theorem 1.1.

Recall, that we say that two subgraphs of a meander

$$\mathcal{M}(\,\cdot \mid \cdot\,) = (\mathrm{V}^{\mathcal{M}}, \mathrm{A}^{\mathcal{M}})$$

are disjoint if their sets of vertices are disjoint. For brevity, we omit the entry $(\,\cdot \mid \cdot\,)$.

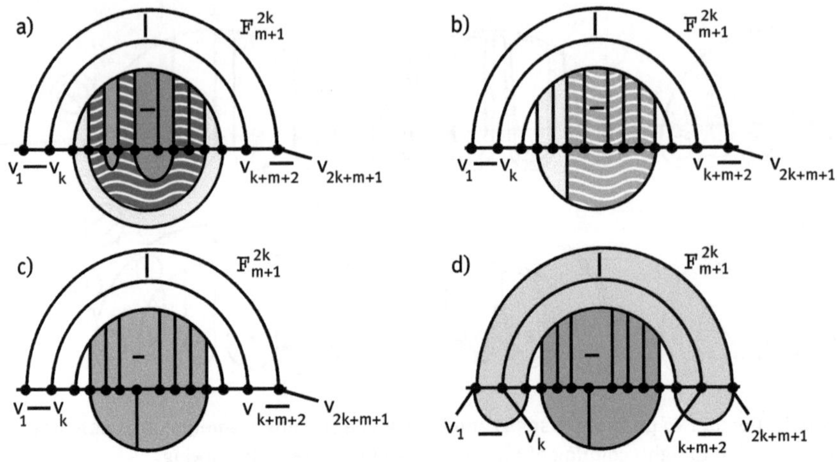

Fig. 2.7: Colored chambers; cases a)–d) discussed below.

Theorem 2.1. *Let $\mathcal{M}$ be a meander and $\mathcal{M}^{coll}$ its collapsed meander. Let $\mathcal{M}_1^{coll}, \ldots, \mathcal{M}_M^{coll}$ be disjoint subgraphs of $\mathcal{M}^{coll}$. Let $K \in \{1, \ldots, M\}$ be the number of disjoint subgraphs of $\mathcal{M}^{coll}$ with at least one chamber. Let $\mathrm{Ch}_i(\mathcal{M}^{coll})$ be the total number of chambers of the ith - disjoint subgraph of the collapsed meander. Then the number of connected components of a meander is given by:*

$$Z(\mathcal{M}) = \left[\sum_{i=1}^{K} \mathrm{Ch}_i(\mathcal{M}^{coll}) + K \right] + P(\mathcal{M}^{coll}) + 2C(\mathcal{M}^{coll}). \tag{2.3}$$

Proof. The proof splits into two cases:
Case a): There are no chambers, and, Case b): There are chambers. To Case a): All connected components of $\mathcal{M}$ are bounded curves in the plane. Therefore, we may color the interior regions of $\mathcal{M}$ and obtain two types: either a bounded ring or a simply-connected domain. By zooming out, the bounded ring domains are cycles and the simply-connected domains are *open* forks, paths or isolated points. By an *open* $(2k, m+1)$-fork we mean that all vertices

$$v_{k+1}, \ldots, v_{k+m}$$

are semi-isolated, i.e., that there are no further arcs emanating from them. Hence, in this case, the collapse of a meander produces cycles, paths including the isolated points, similar to the case of seaweed meanders whose blocks consist of one single level. Since we are not producing any chambers the first part in the sum of (2.3) is zero, i.e.,

$$Z(\mathcal{M}) = P(\mathcal{M}^{coll}) + 2C(\mathcal{M}^{coll}),$$

which is in correspondence with Theorem 1.1.

To Case b): We color $\mathcal{M}$. Here, we may obtain a deformed colored ring domain containing n holes. By zooming out, this corresponds to the case in which n chambers are produced. Since the boundary and each hole of this deformed ring domain correspond to a connected component we must count each hole and the boundary. Therefore, each time we enter a new disjoint subgraph we have to count the number of chambers plus the boundary. Since this is the only one additional difference to *Case a)*, we obtain the desired claim. □

Remarks 2.2.

a) *In chambers there may be other components included such as cycles, paths, isolated points and chambers.*

b) *The above theorem allows us to derive sufficient condition for meanders to be connected.*

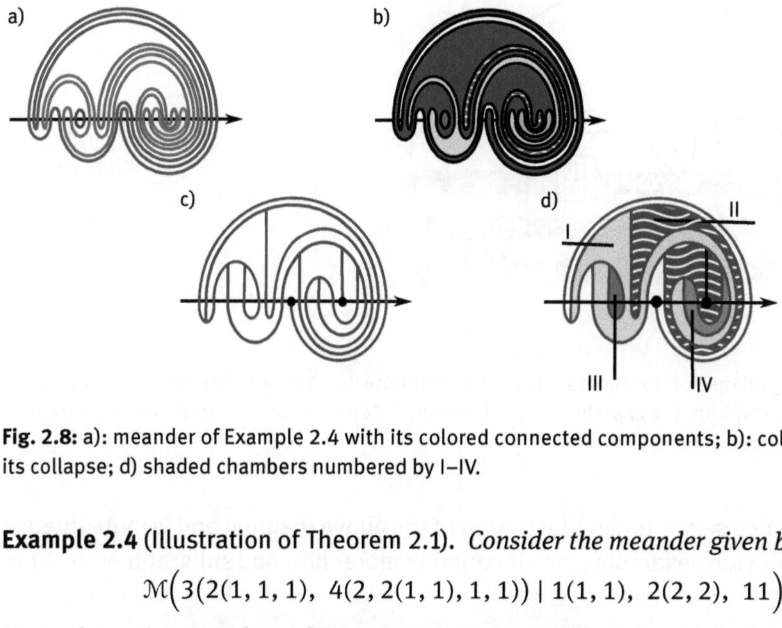

Fig. 2.8: a): meander of Example 2.4 with its colored connected components; b): colored regions; c): its collapse; d) shaded chambers numbered by I–IV.

Example 2.4 (Illustration of Theorem 2.1). *Consider the meander given by*

$$\mathcal{M}\big(3(2(1,1,1),\ 4(2,2(1,1),1,1))\mid 1(1,1),\ 2(2,2),\ 11\big).$$

By applying the procedure of Construction 4 on each block we obtain its collapse

$$\mathcal{M}^{coll}\Big(\Big|^{2}_{3}\Big(\Big|^{0}_{4},\ \Big|^{2}_{5}\big(1,\Big|^{0}_{3},0\big)\Big)\ \Big|\ \Big|^{0}_{3},\ 2(2,2),\ 11\Big).$$

In $\mathcal{M}^{coll}$ there are 4 chambers in only one subgraph, see Figure 2.8. By Theorem 2.1 the number of connected components of $\mathcal{M}$ is

$$Z(\mathcal{M}) = \sum_{i=1}^{1} \mathrm{Ch}_i\big(\mathcal{M}^{coll}\big) + 1 = 5.$$

Indeed, the number of connected components of $\mathcal{M}$ is 5, which is confirmed by Figure 2.8.

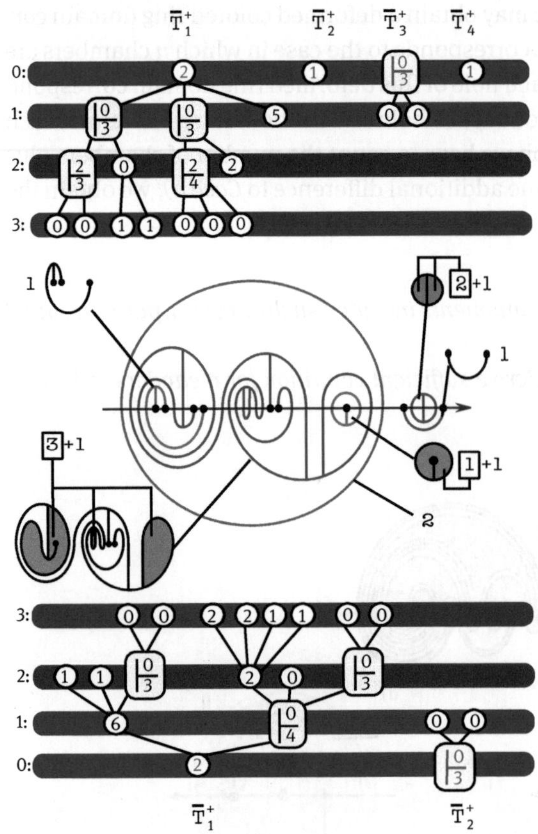

Fig. 2.9: Middle: collapse of the meander from p. 32 with the number of connected components of each disjoint subgraph; the chambers are colored blue. Top and bottom: ordered rooted trees associated to collapsed blocks.

Next example also serves for an illustration of the above theorem and is more interesting than the previous example since it contains more than one subgraph with chambers.

Example 2.5. *Let us consider once again the collapsed meander*

$$\mathcal{M}^{coll}\Big(2\big(\vdash\tfrac{0}{3}\big(\vdash\tfrac{2}{3},1,1,\vdash\tfrac{0}{3}\big(\vdash\tfrac{2}{4},2\big),\ 5\big),1,\vdash\tfrac{0}{3},1\;\Big|\;2\Big(6\big(1,1,\vdash\tfrac{0}{3}\big),\vdash\tfrac{0}{4}\big(2\big(2,2,1,1\big),0,\vdash\tfrac{0}{3}\big)\Big),\vdash\tfrac{2}{3}\Big)$$

from the beginning of the previous section, shown in Figure 2.9. There are seven disjoint subgraphs. Three subgraphs contain chambers. By Adding the numbers depicted in Figure 2.9, we get 13 *connected components which is in correspondence with the number of connected components of the original meander.*

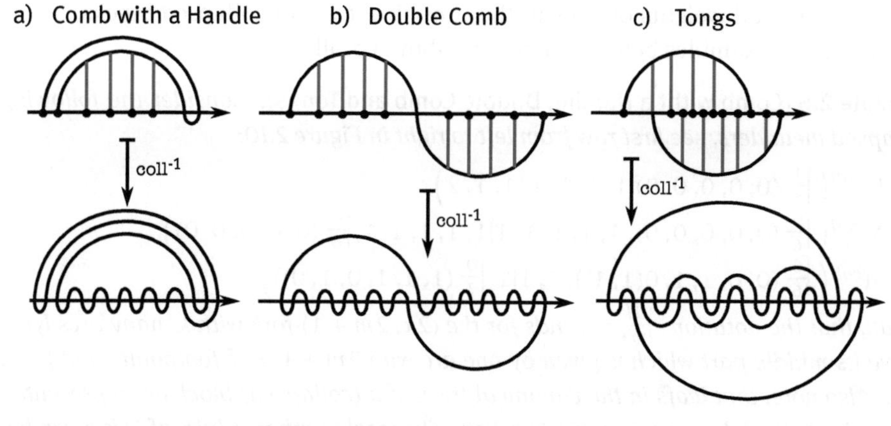

Fig. 2.10: Last row: a) comb with a handle; b) double comb; c) tongs and its collapsed forms in the first row. The bars of the forks are colored red.

2.2 Results and Examples for Connected Meanders

By Theorem 2.1 we have the following corollary:

Corollary 2.1. *The number of connected components of a meander is greater than one, if and only if after its collapse the number of chambers and the number of cycles is not zero or the number of paths is bigger than one. In other words, a meander is disconnected, if and only if after its collapse it contains chambers, cycles or more than one path.*

Therefore, in order to obtain connected meanders we need to exclude the existence of chambers, cycles and multiple paths. The following sufficient condition for connectedness of meanders holds:

Corollary 2.2. *Let $\mathcal{M}$ be a meander given by two compositions*

$$A_i,\ i = 1, \dots, k \text{ and } B_j,\ j = 1, \dots, r,$$

where A_i, B_j are bracket expressions in the Notation 1. Let $\mathcal{M}^{coll}$ be its corresponding collapsed meander. Let

$$\mathrm{F} := \{\mathcal{F}^{2k_i}_{m_i+1} \mid i = 1, \dots, M\}$$

be the set of forks which are in $\mathcal{M}^{coll}$. Let $b_i := m_i - 1$ be the number of bars of the ith-fork and $\mathtt{Odd}(\mathcal{M}^{coll})$ be the number of leafs with odd weights of all canonical trees of $\mathcal{M}^{coll}$. If $\mathcal{M}$ is connected, then

$$\sum_{i=1}^{M} b_i + 2 = \mathtt{Odd}(\mathcal{M}^{coll}).$$

The above Corollary says that if a meander is connected then the sum of all numbers of entries inside the brackets with $\vdots$ in front minus one each plus two equals to the number of all odd weights of leafs in the canonical trees of collapsed blocks. Let us consider some examples before proving the above corollary.

Example 2.6 (Comb witht a Handle, Double Comb and Tongs). *Consider the following collapsed meanders, see first row from left to right in Figure 2.10:*

1) $\mathcal{M}^{coll}\Big(\big|\frac{1}{6}\,(0,0,0,0,0)|1,1,1,1,1,1,2\Big)$,
2) $\mathcal{M}^{coll}\Big(\big|\frac{0}{6}\,(0,0,0,0,0),1,1,1,1,1|1,1,1,1,1,\big|\frac{0}{6}\,(0,0,0,0,0)\Big)$,
3) $\mathcal{M}^{coll}\Big(\big|\frac{0}{6}\,(0,0,1,1,0(1,1)),1,1|1,\big|\frac{0}{7}\,(1,1,1,0,1,0))\Big)$.

Recall, that the notation $\big|\frac{2k}{2m+1}$ *stands for the* $(2k, 2m+1)$*-fork with* k *many arcs lying above its middle part which is given by one arc with* $2m+1 =: f$ *footpoints and* $f-2$ *bars. Also note, that leafs in the canonical trees of a (collapsed) block are represented by numbers which have no opening brackets. The total number of bars of 1) is given by*

$$6-2=4 \quad and \quad 4+2=6,$$

which is the number of all odd numbers represented by ones without opening brackets. Similarly, in 2) the total number of bars is

$$2(6-2)=8 \quad and \quad 8+2=10$$

equals the number of all ones without opening brackets. In 3), we have

$$6-2+7-2=9$$

bars and 11 odd numbers without opening brackets. Indeed, all non-collapsed meanders corresponding to 1), 2) and 3) are connected which is demonstrated by applying the inverse of the collapse map, see the last row in Figure 2.10.

Geometrically, a comb with a handle may only be duplicated and attached after being mirrored vertically in order to produce a connected meander. This idea is apparent in terms of its collapse. In contrast to the comb with a handle, a double comb may be duplicated, say N times, and then attached in such a way that in its collapsed form the right end-vertex is glued with the start-vertex of the next collapsed double comb, see Figure 2.11. In general, this procedure is always possible whenever we have open forks which may be attached to other forks in such a way that no chambers emerge, meaning that the forks stay open after the gluing. In other words, we are not allowed to close up the red bars or any regions lying above the arcs containing the bars when we glue forks with each other. For an example consider the gluing of the double comb, tongs and comb with a handle in Figure 2.11, b).

Proof. [Corollary 2.2] It suffices to consider two cases. Namely, in $\mathcal{M}^{coll}$

1) there are no forks, *or*
2) there are M forks occurring in $\mathcal{M}^{coll}$.

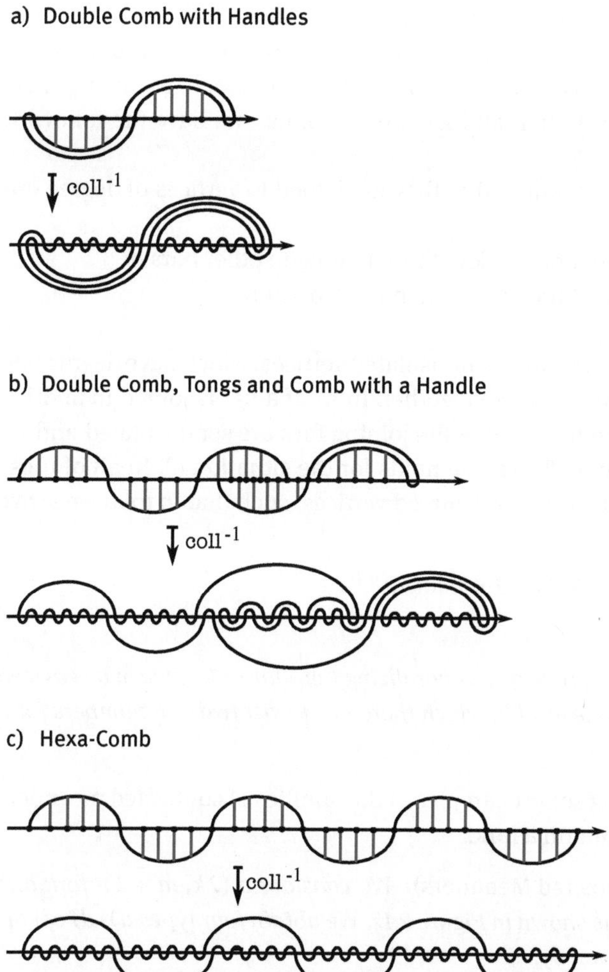

Fig. 2.11: Gluing of the forks in order to produce new connected meanders.

To Case 1): If there are no forks, i.e.,

$$\sum_{i=1}^{M} b_i = 0,$$

then any collapsed meander of a connected meander must be one single path. Since each path must have two semi-isolated vertices, i.e., vertices with degree one, there must be two odd numbers occurring as a weight of a leaf in the corresponding canonical tree of a collapsed block or of two different blocks, regardless whether in $\mathbb{E}^+$ or $\mathbb{E}^-$ or both $\mathbb{E}^+$ and $\mathbb{E}^-$. Hence,

$$\mathtt{Odd}(a_k) = 2.$$

To Case 2): If there are M many $(2k_i, m_i+1)$-forks $\mathrm{F}^{2k_i}_{m_i+1}$ in the collapse of a connected meander, there are no chambers. Otherwise, the number of connected components would be greater than one, see Theorem 2.1. There are no chambers occurring when all forks are joined in such a way that all forks are open, i.e., we have the following three situations:

a) all bars are semi-isolated, meaning that they are joined to vertices of degree one *or*

b) a bar is joined to another bar of a fork without joining to other bars *or*

c) a bar is joined to an arc of a fork without joining to other bars.

In situation a) we must have two further semi-isolated vertices, which have degree one, corresponding to the arc joining M forks together. In b), if a bar is joined to another bar two vertices corresponding to an arc of the joining fork are semi-isolated and we must count as in the situation a). The same holds for the situation c). In all of these situations we must add two further semi-isolated vertices, such that in total we arrive at

$$\sum_{i=1}^{M} b_i + 2 = \mathrm{Odd}(\mathcal{M}^{coll}).$$

□

Remark. *Note, that the above corollary is generalizing Corollary 1.1, since in a seaweed meander all blocks consist of one level in which there must exist two odd numbers for a single component.*

We end this section by showing some examples of the families of connected meanders whose collapsed meanders contain a fork.

Example 2.7 (Families of Connected Meanders). *We consider a $(2k, m+1)$- fork and wind the corresponding bars as shown in Figure 2.12. We obtain four types a)–d) of connected collapsed meanders given by:*

$$a): \quad \mathcal{M}^{coll}\Big(\vdash^{0}_{m+1} (a_1, \dots, a_m), K \mid a_1+1, a_2+1, \dots, a_m+1, K+1 \Big),$$

$$b): \quad \mathcal{M}^{coll}\Big(\vdash^{0}_{m+1} (a_1, \dots, a_{m-1}, 0(a^{2,m}, a^{2,m+1})) \mid a_1+1, \dots, a_{m-1}+1, a^{2,m}+1, a^{2,m+1}+1 \Big),$$

$$c): \quad \mathcal{M}^{coll}\Big(\vdash^{0}_{m+1} (0(a^{2,1}, a^{2,2}), \dots, 0(a^{2,m}, a^{2,m+1})) \mid a^{2,1}+1, a^{2,2}+1, \dots, a^{2,m}+1, a^{2,m+1}+1 \Big),$$

$$d): \quad \mathcal{M}^{coll}\Big(\vdash^{2k}_{m+1} (a_1, \dots, a_m) \mid 1, \underbrace{2, \dots, 2}_{\lceil \frac{k}{2} \rceil -1}, a_1+1, \dots, a_m+1, \underbrace{2, \dots, 2}_{\lceil \frac{k}{2} \rceil} \Big).$$

The corresponding connected meanders $\mathcal{M}$ may be derived from the collapsed connected meanders, i.e.,

$$a): \quad \mathcal{M}(1(a_1, \dots, a_m), K \mid a_1+1, a_2+1, \dots, a_m+1, K+1)),$$

$$b): \quad \mathcal{M}(1(a_1, \dots, a_{m-1}, 0(a^{2,m}, a^{2,m+1})) \mid a_1+1, \dots, a_{m-1}+1, a^{2,m}+1, a^{2,m+1}+1),$$

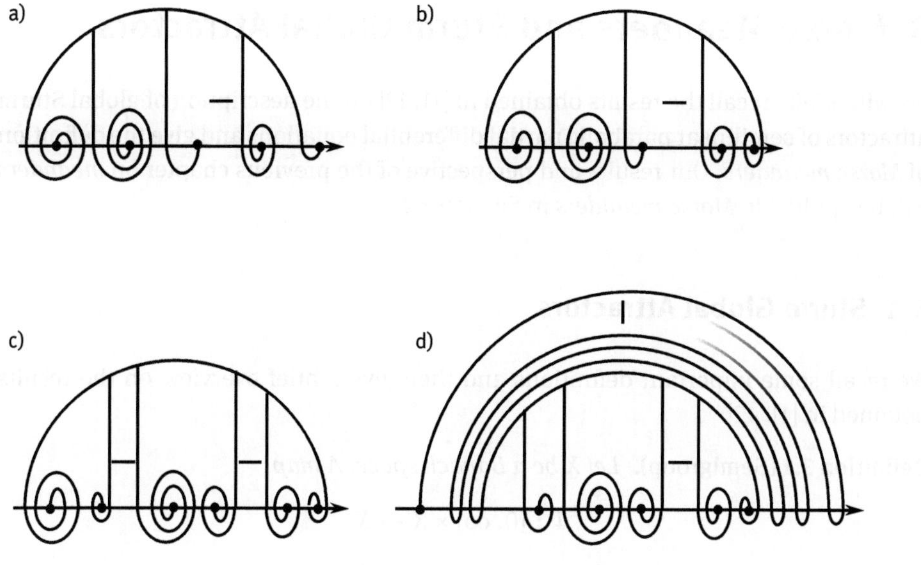

Fig. 2.12: Families of collapsed connected meanders of Type a)-d) obtained from considering forks.

$$c):\quad \mathcal{M}(\,1(0(a^{2,1},a^{2,2}),\dots,0(a^{2,m},a^{2,m+1}))\;|a^{2,1}+1,a^{2,2}+1,\dots,a^{2,m}+1,a^{2,m+1}+1),$$

$$d):\quad \mathcal{M}^{coll}(2k+1(a_1,\dots,a_m)\mid 1,\underbrace{2,\dots,2}_{\lceil\frac{k}{2}\rceil-1},a_1+1,\dots,a_m+1,\underbrace{2,\dots,2}_{\lceil\frac{k}{2}\rceil}).$$

A combination of the four different types produces new types of connected families.

3 Morse Meanders and Sturm Global Attractors

We will briefly recall the results obtained in [11, 12] on the description of global Sturm attractors of semilinear parabolic partial differential equations and give the definition of *Morse meanders*. Our results and perspective of the previous chapter on *meanders* will be applied to *Morse meanders* in Section 3.2.

3.1 Sturm Global Attractors

We recall some important definitions and then give a brief overview on the results obtained in [11, 12]:

Definition 3.1 (Semigroup). *Let X be a Banach space. A map*

$$\Phi : [0, \infty] \times X \to X$$

is called compact strongly continuous semigroup, *if*

$$\Phi : (t, u_0) \mapsto \Phi_t(u_0)$$

is continuous and $\Phi_t(B)$ is relatively compact for any bounded set $B \subseteq X$ such that the semigroup properties hold, i.e.,

$$\begin{aligned} &i) \qquad \Phi^0 = id \\ &ii) \quad \Phi^t \circ \Phi^s = \Phi^{t+s}, \quad \forall t, s \geq 0. \end{aligned}$$

Following [11, 12] we consider a scalar semilinear parabolic partial differential equation with Neumann boundary condition

$$\text{(PDE), (N)} \begin{cases} u_t = a(x)u_{xx} + f(x, u, u_x), \ 0 < x < 1 \\ u_x = 0, \ x = 0 \textit{ or } x = 1. \end{cases} \tag{3.1}$$

The diffusion coefficient $a(x) \geq c_0 > 0$ and the nonlinearity f are assumed to be twice continuously differentiable. In addition, the diffusion coefficient $a(x)$ is assumed to be uniformly positive. Let $X \subset W^{2,2}([0, 1], \mathbb{R})$ be the Sobolev space of function v with square integrable second x-derivative v_{xx} and vanishing v_x at $x = 0$ and $x = 1$. The Equation (3.1) defines a local C^1-semiflow

$$\begin{aligned} u : \mathbb{R} \times X &\to X, \\ (t, u_0) &\mapsto u_t(\cdot) = u(t, \cdot) \end{aligned}$$

on the Sobolev space X, cf. [17]. To the Equation (3.1) one can write down

$$\text{(ODE), (N)} \begin{cases} 0 = a(x)v_{xx} + f(x, v, v_x), \ 0 < x < 1 \\ v_x = 0, \ x = 0 \textit{ or } x = 1. \end{cases} \tag{3.2}$$

DOI 10.1515/9783110533026-004

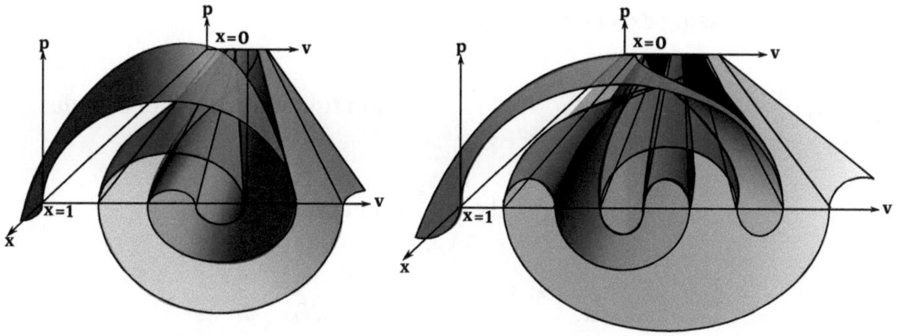

Fig. 3.1: Examples of shooting surfaces.

Definition 3.2 (Shooting Curve, Set of Equilibria). *The image at $x = 1$ of all solutions $(v(x), v_x(x))$ starting from the v-axis $\{x = 0\}$ is called the shooting curve and is denoted by S. The set of equilibria $\mathcal{E}$ evaluated at $x = 1$ is given by the intersection of the shooting curve with the v-axis. In symbols,*

$$\mathcal{E}\big|_{x=1} = \mathrm{S} \cap \{v_x = 0\}.$$

Recall from [15] the definition of a shooting surface, see Figure 3.1 for some examples:

$$M_f := \{(v, p, x) \in \mathbb{R}^2 \times \mathbb{R}_+ \mid v = v(x, a), p = (v_x, a), a \in D_f\}.$$

Explicitly, the shooting curve is given by

$$\mathrm{S} : a \mapsto (v, p)(a, x = 1),$$

where $(v, p)(a, x)$ denotes the solution with initial conditions

$$v = a, p = 0 \text{ at } x = 0$$

of the (ODE)

$$\begin{aligned} v_x &= p \\ p_x &= -f(x, v, p). \end{aligned}$$

Note, that if the intersections of S with the v-axis are strict the shooting curve S is a *meander*. For the Equation (3.1) we introduce the Lyapunow function

$$\mathrm{V} : t \mapsto \mathrm{V}(u(t, \cdot)) := \int_0^1 \left(\frac{1}{2}(u_x)^2 - F(x, u(t, \cdot)) \right) \mathrm{d}x.$$

Note, that V is decreasing except at equilibria, since

$$
\begin{aligned}
\frac{d}{dt}\mathrm{V}(u(t,\cdot)) &= \int_0^1 (u_x u_{xt} - f u_t)\,\mathrm{d}x && \Big|\ \text{chain rule, fund. thm. of calculus} \\
&= [u_x u_t]_0^1 - \int_0^1 (u_{xx} + f) u_t\,\mathrm{d}x && \Big|\ \text{integration by parts} \\
&= -\int_0^1 (u_t)^2 < 0\,\mathrm{d}x. && \Big|\ (PDE), (N)
\end{aligned}
$$

Equilibria v are critical points of V, i.e.,

$$\mathrm{grad}_v \mathrm{V}(v) = 0.$$

Bounded solutions $u(t,\cdot) \in X, t \geq 0$ tend to equilibria for $t \to \infty$. By linearizing Equation 3.1 at equilibrium $v = v(x)$ we obtain the Sturm–Liouville Problem (SLP) with Neumann boundary condition:

$$
\text{(SLP), (N)} \quad \begin{cases} \mu u_t = u_{xx} + f_p(x, v(x), v_x(x)) + f_u(x, v(x), v_x(x)), & 0 < x < 1 \\ u_x = 0, \quad x = 0 \text{ or } x = 1. \end{cases} \tag{3.3}
$$

Definition 3.3 (Hyperbolic Equilibrium, Morse Index). *An equilibrium v is called* hyperbolic, *if* $\mu = 0$ *of* (3.3) *is not an eigenvalue. For v hyperbolic we call the number of strictly positive eigenvalues of* (3.3) *the* Morse index *or the* unstable dimension *and denote it by*

$$\mathrm{i}(v) := |\{\mu_k > 0\}|, \text{ i.e., } \mu_0 > \cdots > \mu_{\mathrm{i}(v)-1} > 0 > \mu_{\mathrm{i}(v)} > \cdots .$$

Lemma 3.1 (Shooting and Hyperbolicity). *An equilibrium v, alias an intersection of the (ODE) shooting curve* S *with the v-axis is hyperbolic if and only if the intersection is transverse, i.e., not tangent to the v-axis.*

Proof. Lecture notes in infinite-dimensional dynamical systems. □

Following [11, 12], for the description of the global attractor $\mathcal{A}$ of (3.1), one needs to impose further assumptions on the nonlinearity f and the equilibria of (3.1), namely

i) f is *dissipative*, i.e.,

$$f(x,u,0)\cdot u < 0$$

for $|u|$ large enough, uniformly in x and

$$|f(x,u,p)| < c_1(u) + c_2(u)|p|^{\beta}$$

with $\beta < 2$ and c_1, c_2 continuous functions,

ii) all equilibria are hyperbolic.

By dissipativeness of f, there exists the global attractor $\mathscr{A}_f$ corresponding to the dynamical system given by the Equation (3.1). Due to the results of [22], the global attractor $\mathscr{A}_f$ is the maximal compact invariant subset of X and attracts all bounded sets. Let $v, w \in \mathscr{E}$ be two equilibria and let $C(v, w)$ denote the heteroclinic connecting orbits $u(t, \cdot)$ which are defined for all real times t and converge to v, w as $t \to -\infty, t \to +\infty$. There is the following decomposition of global attractors of (3.1) by the following theorem which is due to a gradient-like structure, cf. [23], [32]:

Theorem 3.1. *The global attractor $\mathscr{A}_f$ is decomposed into the union of the equilibria and the heteroclinic connections between them. In symbols,*

$$\mathscr{A}_f = \mathscr{E} \cup \bigcup_{v,w \in \mathscr{E}} C(v, w).$$

By the above theorem, for describing $\mathscr{A}_f$ it suffices to understand the heteroclinic connecting orbits $C(v, w)$ once the set of equilibria $\mathscr{E}$ is known. For determining the set of equilibria it is sufficient to solve an ODE problem, whereas the description of $C(v, w)$ is a PDE problem. We will return to a description of $\mathscr{A}_f$ after recalling Morse meanders in the next section by following [9].

3.1.1 CW Complexes: Sturm Global Attractors

In the following and for the convenience of the reader we give a brief account of the recent results from [14]. The following definitions are from [20]. Let

$$B^m = \{v \in \mathbb{R}^m \mid \| v \| \leq 1\}$$

be a m-dimensional closed unit ball. An open *m-cell* is a topological space homeomorphic to the interior of a ball $\mathtt{Int}(B^m)$. The original definition of a *CW complex* by J. H. C. Whitehead reads as follows, cf. [28], cf. [20]:

Definition 3.4 (CW Complex by J. H. C. Whitehead). *Let X be a Hausdorff topological space, and assume that it is represented as a disjoint union of open cells c_β of dimension m. Then, the pair $(X, \{c_\beta\}_\beta)$ is called a* CW complex *if the following two conditions hold:*

1) *For any β there exists a continuous map*

$$f_\beta : B^m \to X$$

   such that
   i) *the restriction of f_β to $\mathtt{Int}(B^m)$ is a homeomorphism onto c_β;*
   ii) *$f_\beta(\partial B^m)$ is a subset of a union of finitely many cells of dimension less than m.*
2) *The subset $A \subseteq X$ is closed in X if and only if*

$$A \cap \overline{c}_\beta$$

   is closed for any β.

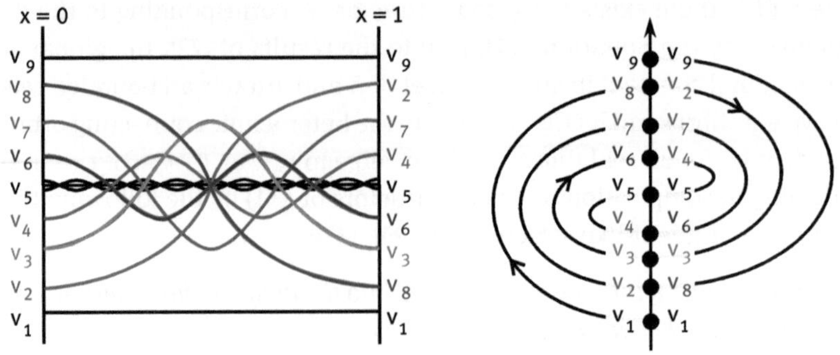

Fig. 3.2: Equilibrium spaghetti and the shooting curve for the Chafee–Infante Problem with nine equilibria.

Definition 3.5 (Regular CW Complex). *A CW complex is called* regular *if for each cell c_β the restriction of the characteristic map*

$$f_\beta : \partial B_\beta \to f_\beta(\partial B_\beta)$$

is a homeomorphism.

In [14], the following theorem is proven:

Theorem 3.2. *Let $\mathscr{A} = \mathscr{A}_f$ be a Sturm global attractor of PDE* (3.1) *and assume all equilibria $v_1, \ldots, v_N$ are hyperbolic. Then,*

$$\mathscr{A}_f = \bigcup_{j=1}^{N} W^u(v_j) \tag{3.4}$$

is a regular dynamic complex, i.e., the dynamic decomposition (3.4) *is a finite regular CW complex with (open) cells given by the unstable manifolds $W^u(v_j)$ of the equilibria v_j.*

3.2 Morse Meanders

Let v_j, $j = 1, \ldots, N$ denote the solutions of (3.2) ordered by their values at $x = 0$:

$$v_1 < v_2 < \cdots < v_N \text{ at } x = 0.$$

Then,

$$\mathscr{E}_f := \{v_1, \ldots, v_N\}$$

is the set of equilibria of (3.1). By reordering the equilibria in $\mathscr{E}_f$ according to the values of v_j we obtain a permutation $\pi_f \in S_N$

$$v_{\pi_f(1)} < v_{\pi_f(2)} < \cdots < v_{\pi_f(N)} \text{ at } x = 1.$$

This permutation π_f is the meander permutation corresponding to the shooting meander S. The reordering of equilibria at $x = 1$ after the shooting causes the so called *equilibrium spaghetti*, for its illustration see Figure 3.2. Originally, G. Fusco and C. Rocha introduced Sturm permutations in order to describe equilibrium spaghetti, cf. [15]. Let us give an example.

Example 3.1 (Equilibrium Spaghetti and Sturm Permutation: C–I Problem). *In the following, we consider the Chafee–Infante problem (C–I Problem) given by*

$$v_{xx} + v(1 - v^2) = 0$$
$$v_x = 0 \quad \text{at } x = 0 \text{ or } x = 1.$$

The outer periodic orbits rotate slower than the inner ones since

$$\frac{f(v)}{v} = 1 - v^2$$

is monotonically decreasing for $v > 0$. Indeed, v is an equilibrium if and only if $(v, p) \in S$, because the Neumann boundary condition at $x = 0$ implies $p = 0$ and $v = a \in \mathbb{R}$. Moreover, Neumann boundary condition at $x = 1$ holds if and only if S inherits the v-axis $\{v_x = 0\}$. In Figure 3.2, we depict the equilibrium spaghetti and the corresponding shooting curve of the Chafee–Infante Problem for $N = 9$. Note that the Sturm permutation is given by $\pi_{CI} = (28)(46)$ in cycle notation.

We briefly recall, the results of [15, 2. A class of admissible permutations] by using the same notation. Let γ be a smooth curve in the (v, p)-plane. Let γ coincide with γ^0 the line of equation $v = p$ outside a bounded set. Assume the intersections of γ to be transversal with the v-axis. Let b_i, $i = 1, \dots, n$ be the abscissae of intersections of γ with the v-axis and let b_i be ordered as they appear when γ is described from (∞, ∞) to $(-\infty, -\infty)$. Define the permutation

$$\sigma_\gamma : \{1, \dots, n\} \to \{1, \dots, n\} \text{ by } \sigma_\gamma(j) > \sigma_\gamma(i) \Leftrightarrow b_j < b_i.$$

The following is introduced by the authors C. Rocha and G. Fusco in order to perform the sequence of pitchfork bifurcations in the definition below:

Step 1 (Angle). Define for each point $z \in \gamma$ the angle $\theta(z)$ such that $\theta(z) + \frac{\pi}{4}$ is the angle (clockwise positive) which is swept by a unit tangent vector of γ when γ is described from (∞, ∞) to z.

Step 2 ("Pitchfork Elements "). Let γ^+ be the smoothing of the open polygon sketched in Figure 3.3. The smoothing of the open polygon is achieved by inserting small circular arcs at the vertices. The reflection of γ^+ on the vertical line is denoted by γ^-.

Step 3 (Appropriate Dilation). Denote by $\delta\gamma_b^\pm$ the arc obtained by a translation of b along the v-axis.

We are ready to recall from [15, 2. A class of admissible permutations] the following definition:

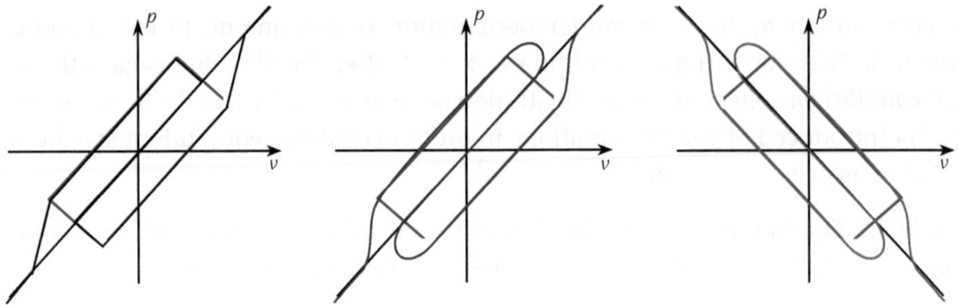

Fig. 3.3: From left to right: open polygon, γ^+, γ^-.

Definition 3.6 (Set of Permutations Π). *A permutation*

$$\sigma_\gamma : \{1, \dots, n\} \to \{1, \dots, n\}$$

is said to be in Π if n is odd and $\sigma = \sigma_\gamma$ for some curve γ such that there is a sequence of curves $\gamma_0, \dots, \gamma_K$, where $K = \frac{n-1}{2}$ with the following properties:

i) $\gamma_0 = \gamma^0$ *and* $\gamma_K = \gamma$,

ii) $\gamma_{i+1}, i = 0, \dots, \frac{n-3}{2}$ *is obtained from γ_i by replacing a segment about one of the intersections $(b, 0)$ of γ_i with the axis $v = 0$ by either $\delta\gamma_b^+$ or $\delta\gamma_b^-$ for some $\delta > 0$.*

iii) *When the replacement in ii) is done with γ^-, the angle function $\theta^i(z)$ corresponding to γ_i satisfies the condition $\theta^i((b, 0)) > 0$.*

The following theorem of [15, 2. A class of admissible permutations, p. 120] guarantees that for permutations corresponding to curves constructed as in the above definition there exists a non-linearity f.

Theorem 3.3. *Assume $\sigma : \{1, \dots, n\} \to \{1, \dots, n\}$ and $\sigma \in \Pi$; then there is a smooth function*

$$f : [0, 1] \times \mathbb{R}^2 \to \mathbb{R}$$

such that σ is the permutation σ_f corresponding to the curve γ_f.

From Definition 3.6 and Theorem 4.2, i) proven in Chapter 4, we know that the permutations determining Morse meanders of type I, II, III and IV which are described by the left and right one-shifts belong to the set Π of the admissible permutations. By Theorem 3.3 [15], permutations determining Morse meanders of type I, II, III and IV describe curves γ_f which are shooting curves corresponding to a (PDE) given in (3.1). By combining the theorems recalled in Chapter 5 for the description of $\mathcal{A}_f$ our investigation of connection graphs of type I, II, III and IV determined by Morse meanders of type I, II, III and IV with respect to graph isomorphisms is legitimate and useful for the classification of $\mathcal{A}_f$.

3.2.1 Definition of Morse Meanders

Following [12], we recall the preliminary definitions such as *dissipative meander permutation, Morse indices* and *Morse permutation*. We define *Morse meanders* in the setting of the first chapter which is equivalent to *Morse meanders* given by dissipative Morse meander permutations.

Definition 3.7 (Dissipative, Meander Permutation). *A permutation $\pi \in S_N$ with N odd is called* dissipative *if it satisfies $\pi(1) = 1$ and $\pi(N) = N$. The permutation π is called a* meander *permutation if it arises from a meander curve, i.e., a connected oriented non-self-intersecting curve which intersects transversely a fixed oriented line in several points.*

The *Morse* indices are the dimensions of the corresponding unstable manifolds. In symbols,

$$\mathtt{i}(v_k) = \dim W^u(v_k).$$

By a result in [25], the shooting meander S determines the *Morse* indices of the equilibria $\mathtt{i}(v_k) \in \mathcal{E}_f$ of the (ODE) (3.2). Explicitly, the Morse indices are given by the following formula:

$$\mathtt{i}(v_k) = \sum_{j=1}^{k-1} (-1)^{j+1} \mathtt{sign}(\pi_f^{-1}(j+1) - \pi_f^{-1}(j)), \tag{3.5}$$

where π_f is the meander permutation and an empty sum denotes zero, cf. [11, Proposition 2.1]. Conversely, to an arbitrary permutation $\pi \in S_N$ a *Morse* index vector $(\mathtt{i}(v_k))_{1 \le k \le N}$ is assigned by using (3.5) or the following recursion

$$\mathtt{i}(v_1) = 0$$
$$\mathtt{i}(v_{k+1}) = \mathtt{i}(v_k) + (-1)^{k+1} \mathtt{sign}(\pi^{-1}(k+1) - \pi^{-1}(k)), \; k = 1, \dots, N-1.$$

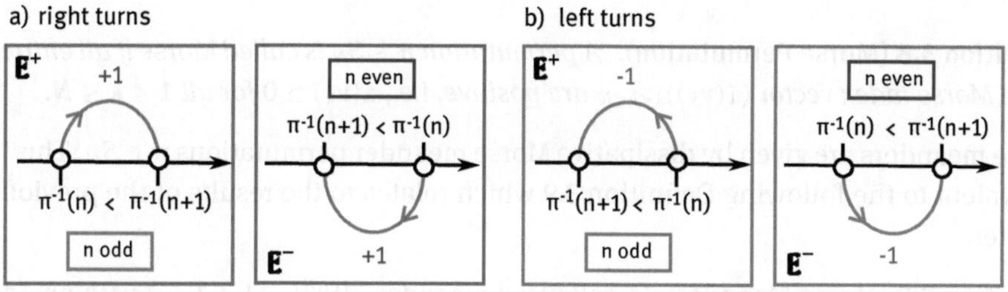

Fig. 3.4: a) Right (green) and b) left (red) turns in the upper and lower half-planes.

There are two types of turns coming into consideration when traveling along a meander, namely left and right turns, see Figure 3.4. It is easy to see by using Formula 3.5,

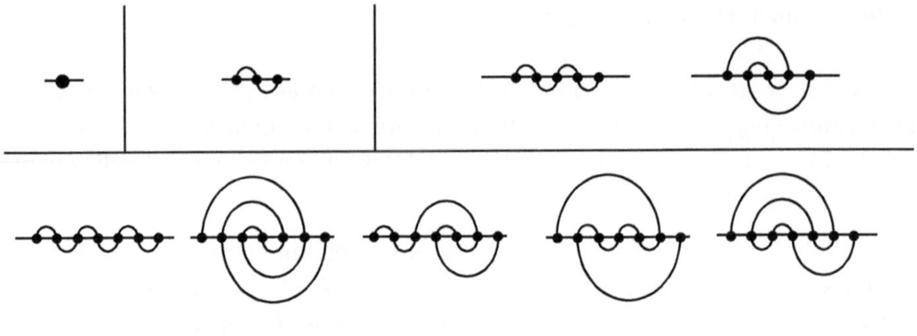

Fig. 3.5: Morse meanders up to seven nodes; in the last row, two Morse meanders which are rotational asymmetric are depicted once.

that a right turn raise a Morse index by one, whereas a left turn decrease it by one. Instead of determining the Morse indices one by one, for a connected meander it suffices to indicate the corresponding alternating sequence of numbers of left and right turns, see Figure 3.4. Note, that for connected meanders from south-west to north-east the sequences start with the number of right turns, whereas for connected meanders from north-west to south-east the initial number is the number of left turns. We denote the number of right turns by $+\mathtt{r}$ and the number of left turns by $-\mathtt{l}$.

Example 3.2. *We consider once again the Chafee–Infante problem for $N = 9$. By using the formula* (3.5) *the Morse indices are:*

$$\mathtt{i}(v_1) = 0,\ \mathtt{i}(v_2) = 1,\ \mathtt{i}(v_3) = 2,\ \mathtt{i}(v_4) = 3,\ \mathtt{i}(v_5) = 4,$$
$$\mathtt{i}(v_6) = 3,\ \mathtt{i}(v_7) = 2,\ \mathtt{i}(v_7) = 1,\ \mathtt{i}(v_9) = 0.$$

Alternatively, the sequence of left and right turns reads as follows $(+4, -4)$. *Note, that the Morse indices may be easily recovered from the sequence of numbers of left and right turns.*

Definition 3.8 (Morse Permutation). *A permutation $\pi \in S_N$ is called* Morse *if all entries in the Morse index vector* $(\mathtt{i}(v_k))_{1\le k\le N}$ *are positive, i.e.,* $\mathtt{i}(v_k) \ge 0$ *for all* $1 \le k \le N$.

Morse meanders are given by dissipative Morse meander permutations $\pi \in S_N$. This is equivalent to the following Definition 3.9 which relates to the results of the previous chapter.

Definition 3.9 (Morse Meander). *A collapsed meander* $\mathcal{M}^{coll}(\,\cdot, 1 \mid 1, \cdot\,)$ *with an odd number of vertices is called a* Morse meander, *if the following conditions are satisfied:*
i) $\mathcal{M}^{coll}(\,\cdot, 1 \mid 1, \cdot\,)$ *is one single path without any forks, i.e there are no branchings.*
ii) *The first vertex v_1 is isolated in $\mathbb{E}^-$ and the last vertex v_N is isolated in $\mathbb{E}^+$, i.e., there is no arc emanating from v_1 in $\mathbb{E}^-$ and no arc emanating from v_N in $\mathbb{E}^+$.*
iii) *The* Morse index vector $(\mathtt{i}(v_k))_{1\le k\le N}$ *is positive, i.e.,* $\mathtt{i}(v_k) \ge 0$ *for all* $1 \le k \le N$.

Remark. *The first condition guarantees that the associated permutation $\pi \in S_N$ is a meander permutation. The second condition stands for the dissipativity of π and is equivalent to the required form $\mathcal{M}^{coll}(\cdot, 1 \mid 1, \cdot)$ in the assumption. Therefore, the second condition in the above definition may be omitted. The last condition guarantees that the permutation is* Morse.

We note that all Morse meanders are given by $\mathcal{M}^{coll}(A, 1 \mid 1, B)$, where A and B stand for compositions of positive even numbers which are the number of nodes in a rainbow or non-rainbow block. This is by Theorem 2.1. There can't be any odd numbers occurring in A or B since the first condition in Definition 3.9 would be violated.

We give some examples for Morse meanders.

Example 3.3. *Let $N = 2K + 1$ be odd.*

a) *The Chafee–Infante meander with nine nodes in Figure 3.2 is given by*

$$\mathcal{M}^{coll}(4, 1 \mid 1, 4).$$

For an arbitrary odd number of nodes N, we end up with $\mathcal{M}^{coll}(2K, 1 \mid 1, 2K)$.

b) *Consider $\pi = id \in S_N$, i.e., the identity permutation. Obviously, it is a dissipative Morse meander permutation. The corresponding Morse meander is given by*

$$\mathcal{M}^{coll}(\underbrace{2, \ldots, 2}_{K}, 1 \mid 1, \underbrace{2, \ldots, 2}_{K}).$$

We end this section by formulating the following open problem.

Problem 3.4. *Let $\mathcal{M}or^{coll}$ denote the set of Morse meanders and $\mathcal{M}^{coll}$ the set of collapsed meanders. Let*

$$\mathcal{M}^{coll}(A, 1 \mid 1, B) \in \mathcal{M}^{coll}$$

be a collapsed meander with compositions A, B containing only even positive integers. Derive the necessary and sufficient conditions on A and B such that $\mathcal{M}^{coll}(A, 1 \mid 1, B)$ is a Morse meander, i.e.,

$$\mathcal{M}^{coll}(A, 1 \mid 1, B) \in \mathcal{M}or^{coll}.$$

In order to solve the above problem, one first needs to derive the necessary condition for connected meanders. The sufficient condition for connected meanders is given by Theorem 2.1. By the Obstruction-Theorem 1 it seems to be a difficult problem if it is possible to derive any single necessary condition for connectedness at all. Secondly, one needs to derive necessary and sufficient condition on the positiveness of the Morse indices associated to the respective vertices of a collapsed meander. The aim of the next two sections is to introduce maps, which will provide us with the possibility of indicating a wide class of Morse meanders.

3.2.2 Raising the Morse Indices

Let us consider two collapsed seaweed meanders with negative Morse indices:

$$\mathcal{S}^{coll}(1, 6 \mid 6, 1)$$

and $\mathcal{S}^{coll}(2, 8, 1 \mid 1, 8, 2)$ (in Figure 3.6, a) and b)).

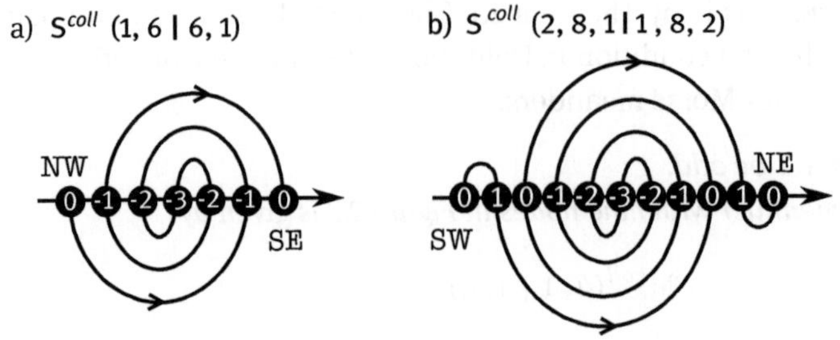

Fig. 3.6: Two meanders with negative Morse indices. In a) the meander starts from north-west with a left turn and in b) from south-west with a right turn. The corresponding Morse indices are written inside the black nodes.

We end up with the following question.

Question 3.5. *Do there exist maps which may be applied to the above meanders*

$$\mathcal{S}^{coll}(1, 6 \mid 6, 1), \ \mathcal{S}^{coll}(2, 8, 1 \mid 1, 8, 2)$$

in such a way that all Morse indices become positive and both meanders are in the set of Morse meanders? If so, do they apply for arbitrary connected meanders with negative Morse indices?

Let $\mathcal{M}^{coll}(A \mid B)$ be a collapsed meander with composition A of positive integers representing the number of nodes in each upper block in the upper-half plane $\mathbb{E}^+$ and composition B of positive integers representing the number of nodes in each lower block in the lower-half plane $\mathbb{E}^-$. Obviously, the following map

$$\begin{aligned} \texttt{mirr} : \mathscr{M}^{coll} &\to \mathscr{M}^{coll} \\ \mathcal{M}^{coll}(A_1, \ldots, A_N \mid B_1, \ldots, B_M) &\mapsto \mathcal{M}^{coll}(\hat{A}_N, \ldots, \hat{A}_1 \mid \hat{B}_M, \ldots, \hat{B}_1). \end{aligned} \tag{3.6}$$

mirrors a collapsed meander on a vertical line by reversing the order of the upper and lower blocks. Note that this is done with respect to each level of the corresponding non-rainbow block, for example if we are given a non-rainbow block with three levels represented by

$$A_k = a_1(a^{2,1}(a^{3,1}, \ldots, a^{3,t}), \ldots, a^{2,p}(a^{3,s}, \ldots, a^{3,m})).$$

The order of the numbers in the bracket is reversed by mirr, i.e., we obtain

$$\hat{A}_k = a_1(a^{2,p}(a^{3,m}, \dots, a^{3,s}), \dots, a^{2,1}(a^{3,t}, \dots, a^{3,1})).$$

For a rainbow block given by some positive A_l, we have $\hat{A}_l = A_l$. The same holds for rainbow and non-rainbow lower blocks. The map mirr applied to our first example $\mathcal{S}^{coll}(1, 6 \mid 6, 1)$ yields

$$\texttt{mirr}(\mathcal{S}^{coll}(1, 6 \mid 6, 1)) = \mathcal{S}^{coll}(6, 1 \mid 1, 6).$$

and already reverses negative Morse indices to positive ones. The reason for this is the interchanging of the first three left turns by right turns which are positive and then are followed by left turns, cf. Proposition 3.1. Hence, if we are given a meander of the form $\mathcal{M}^{coll}(1, A \mid B, 1)$ we apply the mirror map mirr and obtain $\mathcal{M}^{coll}(A, 1 \mid 1, B)$, i.e., a meander from south-west to north-east. We prove the following proposition.

Proposition 3.1. *The mirror map* mirr *applied to a connected meander from south-west to north-east or from north-west to south-east reverses the sequence of left and right turns.*

Proof. Let A_i, $i = 1, \dots, N$ be bracket expressions of positive integers representing the upper blocks and B_j, $j = 1, \dots, M$ be bracket expressions of positive integers representing the lower blocks, such that the total number of arcs in the upper-half plane $\mathbb{E}^+$ equals to the total number of arcs in the lower half-plane $\mathbb{E}^-$. Without loss of generality, let us consider connected meanders from south-west to north-east, i.e., connected meanders of the form $\mathcal{M}^{coll}(A_1, \dots, A_N, 1 \mid 1, B_1, \dots, B_M)$. Since we assume $\mathcal{M}^{coll}(A_1, \dots, A_N, 1 \mid 1, B_1, \dots, B_M)$ to be connected there exists uniquely determined sequence of left and right turns. Let

$$\text{Turn}(\mathcal{M}^{coll}) = (+\texttt{r}_1, -\texttt{l}_1, +\texttt{r}_2, -\texttt{l}_2, \dots, +\texttt{r}_s, -\texttt{l}_s)$$

denote the alternating sequence of numbers of right turns $\texttt{r}_i$, $i = 1, \dots, s$ and left turns $\texttt{l}_i$, $i = 1, \dots, s$. Applying the map mirr on $\mathcal{M}^{coll}(A_1, \dots, A_N, 1 \mid 1, B_1, \dots, B_M)$ is the same as to orient the line from west to east and following the meander from north-east to south-west, i.e., in the reversed order. Therefore, the corresponding sequence of numbers of left and right turns of $\texttt{mirr}(\mathcal{M}^{coll})$ is given by

$$(-\texttt{l}_s, +\texttt{r}_s, \dots, -\texttt{l}_2, +\texttt{r}_2, -\texttt{l}_1, +\texttt{r}_1). \qquad \square$$

The above proposition yields the following corollary.

Corollary 3.1. *Let the entries of the Morse index vector respect the order in which the meander traverses the straight line. The Morse index vector of a connected meander* $\texttt{mirr}(\mathcal{M}^{coll})$ *is obtained from the Morse vector of* $\mathcal{M}^{coll}$ *by reversing the order of its entries and multiplying them by* (-1), *i.e.,*

$$m\Big(\texttt{mirr}\Big(\mathcal{M}^{coll}\Big)\Big)_i = (-1)m\Big(\mathcal{M}^{coll}\Big)_{2k+1-i}.$$

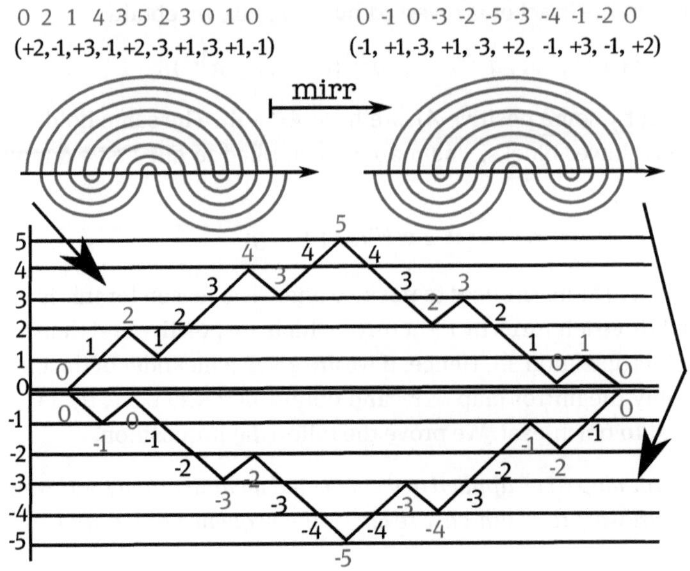

Fig. 3.7: Illustration of the Morse indices of $\mathcal{M}^{coll}(9, 1 \mid 1, 4, 5)$ and $\mathtt{mirr}(\mathcal{M}^{coll}(9, 1 \mid 1, 4, 5))$. The order in which the Morse indices are written down (see the path diagrams below) is with respect to the order in which the meander is traversing the straight line by following its orientation, i.e., from south-west to north-east.

For an illustration of the above results, see Figure 3.7. For connected meanders $\mathcal{M}^{coll}(A, 1 \mid 1, B)$ from south-west to north-east with negative Morse indices we may add sufficiently many right turns in the beginning. Let $\mathrm{i}(v_k)$ be its maximal negative Morse index and

$$\alpha := \max |\mathrm{i}(v_k)|.$$

Then, we have to append α-many right turns above $\mathcal{M}^{coll}(A, 1 \mid 1, B)$ in $\mathbb{E}^+$ and α-many left turns below $\mathcal{M}^{coll}(A, 1 \mid 1, B)$ in $\mathbb{E}^-$. This is realized as follows.

Construction 5 (α-**Suspension**). *Let $v_1, \ldots, v_N$ be the vertices of a connected meander $\mathcal{M}^{coll}(A, 1 \mid 1, B)$ ordered on the straight line $l = \mathbb{E}^+ \cap \mathbb{E}^-$ and $\max|-\mathrm{i}(v_k)| = \alpha$. Add α-many vertices $v_1^*, \ldots, v_\alpha^*$ preceding the vertices of $\mathcal{M}^{coll}(A, 1 \mid 1, B)$ and α-many vertices $v_{N+1}^*, \ldots v_{N+\alpha}$ following after the vertices of $\mathcal{M}^{coll}(A, 1 \mid 1, B)$. We leave the first new vertex v_1^* isolated in $\mathbb{E}^-$ and the last new vertex $v_{N+\alpha}^*$ isolated in $\mathbb{E}^+$, i.e., we do not connect them in the respective half-planes to other vertices. We connect the remaining new vertices by arcs following the rainbow nesting in $\mathbb{E}^+$ and $\mathbb{E}^-$, i.e., in $\mathbb{E}^+$: v_1^* is adjacent to $v_{N+\alpha-1}$, v_2^* is adjacent to $v_{N+\alpha-2}$, …, v_α^* is adjacent to v_N; and in $\mathbb{E}^-$: v_2^* is adjacent to $v_{N+\alpha}^*$, v_3^* is adjacent to $v_{N+\alpha-1}$, …, v_α is adjacent to v_{N+2}^*, v_1 is adjacent to v_{N+1}^*.*

The above construction may be shortened by introducing the following map

$$\begin{aligned} \mathtt{sus}_\alpha : \mathscr{M}^{coll} &\to \mathscr{M}^{coll} \\ \mathcal{M}^{coll}(A, 1 \mid 1, B) &\mapsto \mathcal{M}^{coll}(2\alpha(A), 1 \mid 1, 2\alpha(B)). \end{aligned} \tag{3.7}$$

The map $\mathtt{sus}_\alpha$ raises all Morse indices of the original meander $\mathcal{M}^{coll}(A, 1 \mid 1, B)$ by α since we have appended α-many right turns, see Figure 3.8. Returning to our example in the beginning of the section, we apply $\mathtt{mirr}$ and $\mathtt{sus}_2$ and obtain meanders with positive Morse indices, see Figure 3.9. We summarize the above considerations in order to obtain meanders containing only positive Morse indices.

Construction 6 (Mirror and Suspend). *Let $\mathtt{i}(v_k)$ denote the maximal negative Morse index. We consider two cases.*

a) *Let $\mathcal{M}^{coll}(A, 1 \mid 1, B)$ be a connected collapsed meander with A and B as before. Let $\alpha := \max |\mathtt{i}(v_k)|$. Then, by applying $\mathtt{sus}_\alpha$ to $\mathcal{M}^{coll}(A, 1 \mid 1, B)$, we obtain*

$$\mathcal{M}^{coll}(2\alpha(A), 1 \mid 1, 2\alpha(B))$$

containing only positive Morse indices. In particular, it holds by construction that the new Morse indices corresponding to the vertices of the original meander are obtained by adding α, namely

$$\mathtt{i}(\hat{v}_k) = \mathtt{i}(v_k) + \alpha, \quad k = 1, \ldots, \left|V^{\mathcal{M}^{coll}(A,1|1,B)}\right|.$$

b) *Let $\mathcal{M}^{coll}(1, A \mid B, 1)$ and let $\alpha := \max |\mathtt{i}(v_k)|$. In order to get rid of negative Morse indices we need to apply the map $\mathtt{mirr}$ on $\mathcal{M}^{coll}(1, A \mid B, 1)$ first and then perform the α-suspension by $\mathtt{sus}_\alpha$.*

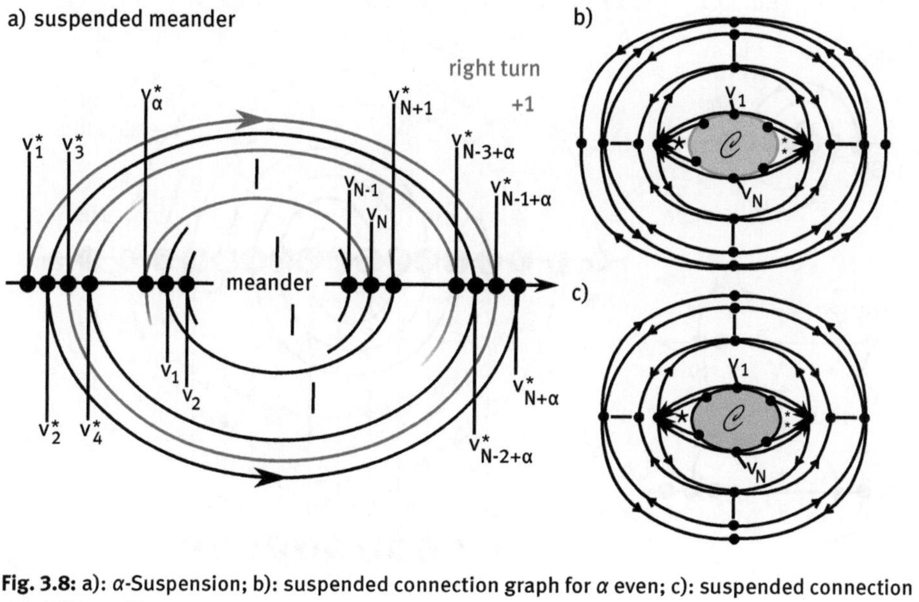

Fig. 3.8: a): α-Suspension; b): suspended connection graph for α even; c): suspended connection graph for α odd.

There is another way to suspend a connected collapsed meander from north-west to south-east

$$\mathcal{M}^{coll}(1, A \mid B, 1)$$

by the map $\mathtt{sus}_\alpha$ without having to mirror it before. Instead, we may append an upper arc to the first vertex in $\mathbb{E}^+$ and a lower arc to the last vertex in $\mathbb{E}^-$ by introducing the map mut

$$\begin{aligned} \mathtt{mut} : \mathscr{M}^{coll} &\to \mathscr{M}^{coll} \\ \mathcal{M}^{coll}(1, A \mid B, 1) &\mapsto \mathcal{M}^{coll}(2, A, 1 \mid 1, B, 2). \end{aligned} \tag{3.8}$$

The abbreviated name of the above map is chosen from the first three letters of latin word "mutare" to *change*. The map mut changes north-west by south-west and south-east by north-east, see Figure 3.9, c). Note, that we only need to apply $\mathtt{sus}_{\alpha-1}$, since we have one additional right turn produced by the map mut, see Figure 3.9, c). It is easy to see that the maps mirr and mut are the two simplest maps which may be applied in order to change the orientations, i.e., north-west by south-west and south-east by north-east of a connected meander $\mathcal{M}^{coll}(1, A \mid B, 1)$.

The Question 3.5 may be answered as follows: In order to raise the Morse indices of a connected collapsed meander we apply, if necessary, the maps mirr or mut and then suspend sufficiently many times by the map $\mathtt{sus}_{(\cdot)}$.

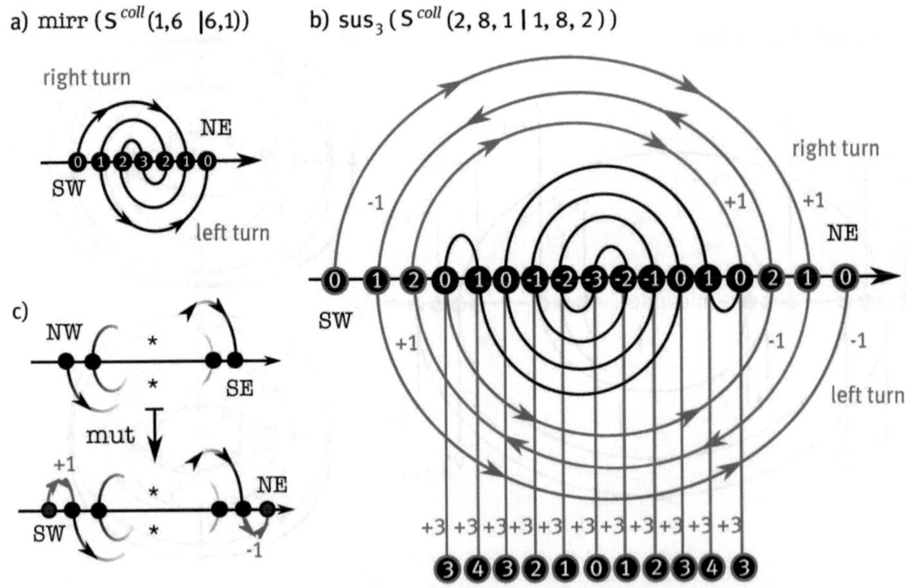

Fig. 3.9: a),b)): Examples on raising negative Morse indices; c): visualization of the map mut.

4 Right and Left One-Shifts

We give a combinatorial description of Morse meanders of *type I, II, III* and *IV* by introducing left and right one-shifts and prove in Theorem 4.2, in i), that they are *pitchforkable*. From [15] by Fusco and Rocha, we know that permutations describing Morse meanders of these types are in the class of admissible permutations. The four different types of Morse meanders will be essential for the further study of the corresponding connection graphs in the next chapter.

4.1 Morse Meanders of Type I

Let $A = A_1, \dots, A_N$ denote the bracket expressions representing upper blocks and let $B = B_1, \dots, B_M$ denote the bracket expressions representing lower blocks in the Notation 1 with respect to arcs. Let n be the total number of upper, respectively lower arcs. We prove that open meanders

$$\mathcal{M}^{coll}(2A, 1 \mid 1, 2A)$$

are Morse meanders. The proof is based on the *right one-shift* applied to closed meanders which is defined as follows. Let v_k, $k = 1, \dots, 2n$, be the vertices corresponding to the upper blocks and

$$v_{k'}, k' = 1', \dots, (2n)',$$

the vertices corresponding to the lower blocks. In the upper half-plane $\mathbb{E}^+$ we fix the upper blocks, whereas in the lower half-plane $\mathbb{E}^-$ we shift all arcs and vertices by one to the right and identify

$$v_k, k = 2, \dots, 2n, \quad \text{with} \quad v_{(k-1)'}, k' = 1', \dots, (2n-1)'.$$

Formally, we introduce the following map:

$$\begin{aligned} \mathtt{shift}_{[1]} : \mathcal{M} &\to \mathcal{M}^{SW \rightsquigarrow NE} \\ \mathcal{M}(A \mid B) &\mapsto \mathtt{shift}_{[1]}(\mathcal{M}(A \mid B)) =: \mathcal{M}(A \mid B[1]). \end{aligned} \tag{4.1}$$

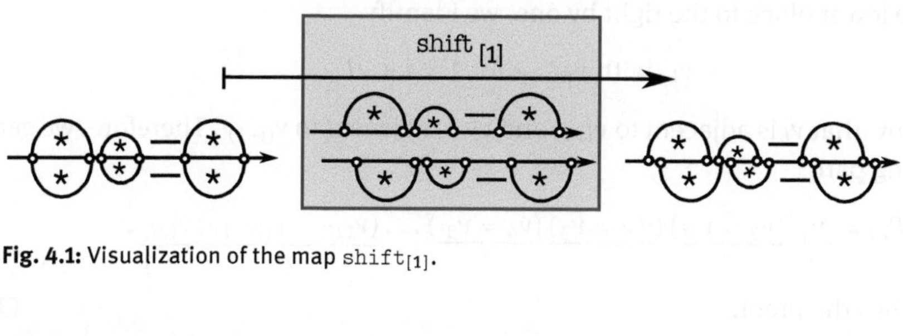

Fig. 4.1: Visualization of the map $\mathtt{shift}_{[1]}$.

DOI 10.1515/9783110533026-005

The image $\mathcal{M}(A \mid B[1])$ is an open or open and closed meander from south-west to north-east which is not necessarily connected for unequal A and B, see the discussion in Section 4.5. When the compositions of lower and upper blocks coincide, we call $\mathcal{M}(A \mid A[1])$ *right one-shifted elements*. Before proving Theorem 4.1, we need the following proposition illustrated in Figure 4.2.

Proposition 4.1.

i) *Let $a > 1$ be the number of arcs of a rainbow block. Then, $\mathcal{M}(a \mid a[1])$ are Morse meanders whose meander permutations are Chafee–Infante, i.e., given by*

$$\pi^a_{CI} = [(2k \quad 2(a-k+1)]_{k=1,\dots,\lfloor \frac{a}{2} \rfloor}$$

in cycle notation.

ii) *Let $\overline{1} := (1, \dots, 1)$ denote l-many arcs ordered on the straight line next to each other. Then, $\mathcal{M}(\overline{1} \mid \overline{1}[1])$ are Morse meanders, whose meander permutations are given by $\pi = id \in S_l$.*

Proof. *i):* Let $v_1, \dots, v_{2a}$ be the vertices of the upper rainbow block $\mathcal{B}^+$ in $\mathbb{E}^+$ with a-many arcs and $v_{1'}, \dots, v_{(2a)'}$ vertices of the lower block $\mathcal{B}^-$ in $\mathbb{E}^-$. Before applying the right one-shift on $\mathcal{M}(a \mid a)$ we get a-many interleaved cycles by identifying

$$v_k \text{ with } v'_k, k = 1, \dots, 2a.$$

By shifting the lower block to the right by one, we identify

$$v_k \text{ with } v_{(k-1)'} \text{ for } 1 \le k \le 2a.$$

Since the lower and upper block are rainbow, i.e., v_k is adjacent to v_{2a-k+1} for all

$$1 \le k \le 2a$$

and $v_{k'}$ is adjacent to $v_{(2a-k+1)'}$ for all $1' \le k' \le (2a)'$, we get the following path:

$$P_{CI} = \underbrace{v_1}_{1}\underbrace{v_{2a}}_{2}\underbrace{v_3}_{3}\underbrace{v_{2a-2}}_{4}\underbrace{v_5}_{5}\underbrace{v_{2a-4}}_{6}\cdots\underbrace{v_{a+1}}_{a+1}\cdots\underbrace{v_6}_{2a-4}\underbrace{v_{2a-3}}_{2a-3}\underbrace{v_4}_{2a-2}\underbrace{v_{2a-1}}_{2a-1}\underbrace{v_2}_{2a}\underbrace{v_{(2a)'}}_{2a+1}.$$

ii): Let $v_1, \dots, v_{2l}$ be the vertices of the upper rainbow block $\mathcal{B}^+$ in $\mathbb{E}^-$ with a-many arcs and $v_{1'}, \dots, v_{(2l)'}$ of the lower block $\mathcal{B}^-$ in $\mathbb{E}^-$. Before applying the one-shift on $\mathcal{B}^-$ to the right we get l-many cycles lying side by side by identifying v_j with v'_j. By shifting the lower block to the right by one, we identify

$$v_k \text{ with } v_{(k-1)'} \text{ for } 1 \le j \le 2l.$$

By $\overline{1}$ we know that v_i is adjacent to v_{i+1}, and $v_{i'}$ is adjacent to $v_{(i+1)'}$. Therefore, we get the following path

$$P_{id} = \underbrace{v_1}_{1}\underbrace{(v_2 = v_{1'})}_{2}\underbrace{(v_{2'} = v_3)}_{3}\underbrace{(v_4 = v_{3'})}_{4}\cdots\underbrace{(v_{(2l)} = v_{(2l-1)'})}_{2l}\underbrace{v_{(2l)'}}_{2l+1},$$

which finishes the proof. □

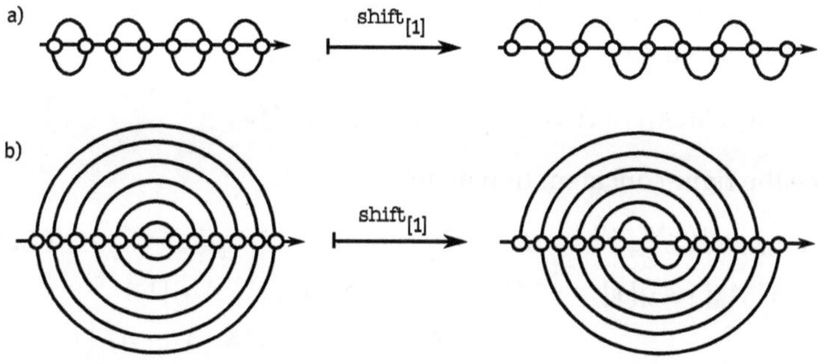

Fig. 4.2: a): Right identity element; b) Chafee–Infante element of order six .

Definition 4.1 (Chafee–Infante Elements, Right Identity Elements). *Let $a > 1$ and*

$$\overline{1} = \underbrace{(1, \ldots, 1)}_{l} .$$

We call $\mathcal{M}(a \mid a[1])$ Chafee–Infante element *of order a and* $\mathcal{M}(\overline{1} \mid \overline{1}[1])$ right identity element.

Note that Chafee–Infante elements of order a have maximal Morse index a. We are ready to formulate and prove the following theorem illustrated in Figure 4.3.

Theorem 4.1. *Let $A = A_1, \ldots, A_M$ be a composition representing upper/ lower blocks. Then, right one-shifted elements*

$$\texttt{shift}_{[1]}(\mathcal{M}(A \mid A)) = \mathcal{M}(A \mid A[1])$$

i) *are connected meanders*[1]
ii) *with positive Morse vectors.*

In other words, right one-shifted elements with same compositions of upper and lower blocks $\texttt{shift}_{[1]}(\mathcal{M}(A \mid A))$ *are Morse meanders.*

Proof. We split the proof into two steps.

Step 1. We show that it suffices to argue for meanders given by one single block $\mathscr{B}$ represented by some bracket expression A_j. We consider

$$\texttt{shift}_{[1]}(\mathcal{M}(A_j \mid A_j)), j = 1, \ldots, M.$$

By definition of the right one-shift map the initial vertex i_{A_j} and the final vertex f_{A_j} are semi-isolated in the image of the right one-shift map, i.e the vertex i_{A_j} has no incident

1 in the sense of Arnol'd's definition, cf. Introduction.

arcs in $\mathbb{E}^-$ and f_{A_j} has no incident arcs in $\mathbb{E}^+$. By identifying the final vertex f_{A_j} with the initial vertex $i_{A_{j+1}}$, we concatenate $\mathtt{shift}_{[1]}(\mathcal{M}(A_j \mid A_j))$ with

$$\mathtt{shift}_{[1]}(\mathcal{M}(A_{j+1} \mid A_{j+1})) \text{ for all } j = 1, \ldots, M.$$

Formally, we introduce the right-concatenation map:

$$\circ_R : (\mathscr{M}^{SW \rightsquigarrow NE})^{\times M} \to \mathscr{M}^{SW \rightsquigarrow NE} \tag{4.2}$$

$$\mathcal{M}(A_1 \mid A_1[1]), \ldots, \mathcal{M}(A_M \mid A_M[1]) \mapsto \mathcal{M}(A_1, \ldots A_M \mid (A_1, \ldots, A_M)[1])$$
$$=: \mathcal{M}(A_1 \mid A_1[1]) \circ_R \ldots \circ_R \mathcal{M}(A_M \mid A_M[1]).$$

By definitions of the maps $\mathtt{shift}_{[1]}$ and $\circ_R$ we have the following correspondence:

$$\mathcal{M}(A_1, \ldots, A_M \mid (A_1, \ldots, A_M)[1]) = \mathcal{M}(A_1 \mid A_1[1]) \circ_R \ldots \circ_R \mathcal{M}(A_M \mid A_M[1]).$$

Since the right-concatenation $\circ_R$ preserves connectedness, it suffices to prove that

$$\mathtt{shift}_{[1]}(\mathcal{M}(A_j \mid A_j))$$

is connected. In each $\mathtt{shift}_{[1]}(\mathcal{M}(A_j \mid A_j))$ the number of left turns coincides with the number of right turns since in the lower and upper half-planes we shift the same configuration of arcs. Therefore, the Morse indices of other elements

$$\mathtt{shift}_{[1]}(\mathcal{M}(A_k \mid A_k)), \ k > j$$

are not influenced by the Morse indices of $\mathtt{shift}_{[1]}(\mathcal{M}(A_j \mid A_j))$. Hence, it suffices to prove that $\mathtt{shift}_{[1]}(\mathcal{M}(A \mid A))$ is Morse, where A represents one single block.

Step 2. Let n be the total number of arcs of A representing one block $\mathscr{B}_A$. We show by strong induction on the total number of arcs of $\mathscr{B}_A$ that a right one-shifted element $\mathtt{shift}_{[1]}(\mathcal{M}(A \mid A))$ is a Morse meander.

Induction start: $n = 1$, i.e., $A = 1$. Then, $\mathtt{shift}_{[1]}(\mathcal{M}(1 \mid 1))$ is the identity element on three vertices which is connected with positive Morse indices

$$\mathrm{i}(v_1) = 0, \mathrm{i}(v_2) = 1, \mathrm{i}(v_3) = 0.$$

Induction assertion: For A representing $k \geq 1$ blocks with n arcs in total a right one-shifted element $\mathtt{shift}_{[1]}(\mathcal{M}(A \mid A))$ is a Morse meander.

Induction step: $[n \curvearrowright n+1]$. Let $\overline{A}$ represent a block with $n+1$ arcs which differ from A by one arc. Since $\mathscr{B}_{\overline{A}}$ is a block we may remove the outermost arc. The resulting arc configuration represented by A consists of k blocks $\mathscr{B}_1, \ldots, \mathscr{B}_k$ represented by $B_1, \ldots, B_k$. Since each block of $\mathscr{B}_j$ has at most n arcs, it follows by induction assertion that the right one-shifted elements

$$\mathtt{shift}_{[1]}(\mathcal{M}(B_j \mid B_j)), \ j = 1, \ldots, k$$

are Morse meanders. From Step 1, we know in particular that

$$\mathtt{shift}_{[1]}(\mathcal{M}(A \mid A)) = \mathtt{shift}_{[1]}(\mathcal{M}(B_1 \mid B_1)) \circ_R \ldots \circ_R \mathtt{shift}_{[1]}(\mathcal{M}(B_k \mid B_k)).$$

To conclude our proof, we consider

$$\begin{aligned}\mathtt{shift}_{[1]}(\mathcal{M}(\overline{A} \mid \overline{A})) &= \mathtt{shift}_{[1]}(\mathcal{M}(1(A) \mid 1(A)) \\ &= \mathtt{sus}_1 \underbrace{(\mathtt{shift}_{[1]}(\mathcal{M}(A \mid A)))}_{\text{Induction assertion}},\end{aligned}$$

where $\mathtt{sus}_1 : \mathscr{M}^{SW \rightsquigarrow NE} \circlearrowleft$ stands for suspending a meander from south-west to north-east. Since suspensions are preserving connectivity and raise the Morse indices of $\mathtt{shift}_{[1]}(\mathcal{M}(A \mid A))$ by one, it follows that $\mathtt{shift}_{[1]}(\mathcal{M}(\overline{A} \mid \overline{A}))$ is a Morse meander, which finishes the proof. □

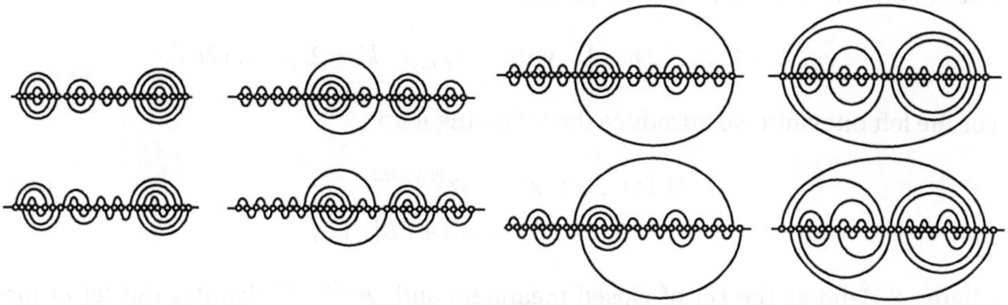

Fig. 4.3: First row: closed meanders $\mathcal{M}(A \mid A)$; second row: examples of Morse meanders of the form $\mathcal{M}(A \mid A[1])$.

Remarks 4.1.

1) *Let n be the total number of arcs of A as in the assumption of the above theorem. The above theorem encodes C_n-many Morse meanders, where $C_n = \frac{2n!}{(n+1)!n!}$ are the Catalan numbers for $n \geq 1$. This is due to the well-known fact, cf. [10], that we have C_n-many possible ways of building blocks. Hence, the Catalan number is a lower bound for the number of Morse meanders.*
2) *In the notation of the previous sections, the above theorem guarantees that*

$$\mathcal{M}^{coll}(2A, 1 \mid 1, 2A)$$

are Morse meanders.

We end this section by introducing the following construction:

Construction 7 (Morse Meanders of Type I). *Let $A = A_1 \ldots, A_N$ represent N-many blocks. We call Morse meanders constructed from right one-shifted elements*

$$\mathtt{shift}_{[1]}(\mathcal{M}(A \mid A)) = \mathcal{M}(A \mid A[1])$$

Morse meanders of type I. *We denote the class of Morse meanders of type I by* $[\mathcal{M}]_{[1]}$.

4.2 Morse Meanders of Type II

In the sequel, we introduce left one-shifts of $\mathcal{M}(A \mid A)$ in the following way. We label the vertices of the upper blocks given by

$$A = A_1, \ldots, A_M \text{ by } v_k, \; k = 1, \ldots, 2n$$

and the vertices corresponding to lower blocks given by

$$B = B_1, \ldots, B_N \text{ by } v'_j, \; j' = 1', \ldots, (2n)'.$$

The left one-shift is introduced by gluing

$$v_k, \; k = 1, \ldots, 2n-1 \quad \text{with} \quad v_{(k+1)'}, \; k' = 2', \ldots, (2n)'.$$

For the left one-shift we introduce the following map:

$$\begin{aligned} \texttt{shift}_{[-1]} : \mathscr{M} &\to \mathscr{M}^{NW \rightsquigarrow SE} \\ \mathcal{M}(A \mid B) &\mapsto \mathcal{M}(A \mid B[-1]), \end{aligned} \tag{4.3}$$

where $\mathscr{M}$ denotes the set of closed meanders and $\mathscr{M}^{NW \rightsquigarrow SE}$ denotes the set of meanders from north-west to south-east which are not necessarily connected.

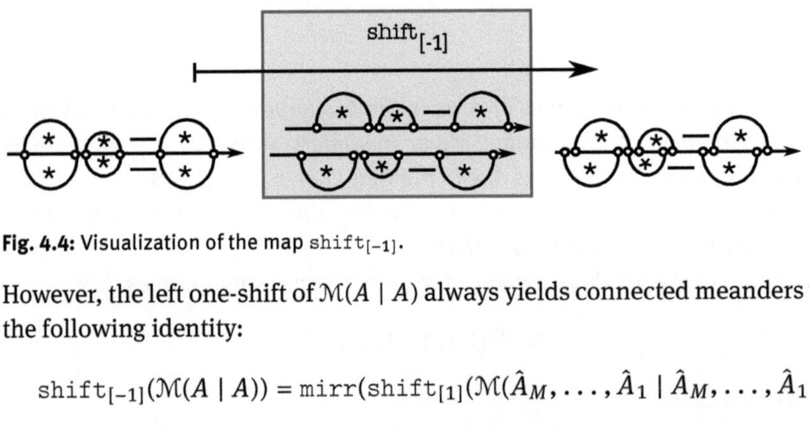

Fig. 4.4: Visualization of the map $\texttt{shift}_{[-1]}$.

However, the left one-shift of $\mathcal{M}(A \mid A)$ always yields connected meanders which is by the following identity:

$$\texttt{shift}_{[-1]}(\mathcal{M}(A \mid A)) = \texttt{mirr}(\texttt{shift}_{[1]}(\mathcal{M}(\hat{A}_M, \ldots, \hat{A}_1 \mid \hat{A}_M, \ldots, \hat{A}_1))), \tag{4.4}$$

where "$\hat{\ }$" reverses the order of the numbers inside the brackets of each A_i for $i = 1, \ldots, M$, compare with the previous chapter. Due to the following result, it is possible to recover Morse meanders by sufficiently many suspensions of

$$\texttt{shift}_{[-1]}(\mathcal{M}(A \mid A)).$$

Definition 4.2 (Left One-Shifted Elements). *Let A be a bracket expression representing a general block with K levels and let T_A be its canonical tree. We call connected meanders* $\mathcal{M}(A \mid A[-1])$ left one-shifted elements.

Proposition 4.2. *Let $A_1, \dots, A_N$ represent blocks and let*

$$\mathcal{M}(A_1, \dots, A_N \mid (A_1, \dots, A_N)[-1])$$

be a left one-shifted element. Then, the maximal negative Morse index of

$$\mathcal{M}(A_1, \dots, A_N \mid (A_1, \dots, A_N)[-1])$$

is given by $\max_{i=1,\dots,N}\{y_i\} =: y_k$*, which is determined as follows:*
1) *Compute the sums over maximal weights along all maximal paths of each canonical tree* T_{A_i}*,* $i = 1, \dots, N$*.*
2) *Consider the greatest sums*

$$y_1, \dots, y_N$$

computed in 1) and pick the maximal y_k*,* $k \in \{1, \dots, N\}$*.*

Proof. The maximal positive Morse index of a right one-shifted element

$$\mathtt{shift}_{[1]}(\mathcal{M}(A_j \mid A_j))$$

is given by the largest sum over all weights along a maximal path of the canonical tree T_{A_j}. This holds by the property of multiple suspensions which raise the Morse indices by a certain value. By using the identity in (4.4) and Corollary 3.1, we obtain the desired claim for the maximal negative Morse index of $\mathtt{shift}_{[-1]}(\mathcal{M}(A \mid A))$. □

By Proposition 4.2 and by using maps $\mathtt{mut}$ and $\mathtt{sus}_{(\cdot)}$ from Chapter 3, Section 3.2.2, we know how to construct Morse meanders out of left one-shifted elements. We define Morse meanders of type II as follows:

Definition 4.3 (Morse Meanders of Type II). *Let* $A_j \neq (1, \dots, 1)$*, which also means that* $A_j \neq 1$*, be an expression representing a general block and let* $y_j \in \mathbb{N}_{>0}$ *be determined as described in Proposition 4.2. We call Morse meanders obtained by using the right concatenation* $\circ_{\mathrm{R}}$*, multiple suspensions* $\mathtt{sus}_{(\cdot)}$ *and the right one-shift applied to closed meanders of the form* $\mathcal{M}(1, A_j \mid A_j, 1)$, Morse meanders of type II.

By the above definition, we have the freedom to choose when to apply multiple suspensions and when to concatenate. Precisely, we describe the class of Morse *meanders of type II*, which splits into five subclasses *A*), *B*), *C*), *D*), *E*), see Figure 4.5, as follows:

Construction 8 (Morse meanders of type II, A), B), C), D), E)). *Let*

$$A_i, i = 1, \dots, M$$

represent blocks with L_i*,* $i = 1, \dots, M$ *levels and* T_{A_i} *be its canonical trees. Furthermore assume,* $A_i \neq (1, \dots, 1)$*, which also means that* $A_i \neq 1$*. Let*

$$\mathcal{M}(A_i \mid A_i[-1])$$

be left one-shifted elements and let y_i *be determined as in Proposition 4.2. In order to construct Morse meanders out of* $\mathcal{M}(A_i \mid A_i[-1])$ *we have the following possibilities:*

A) *We first apply the map* $\texttt{mut}$ *on each of* $\mathcal{M}(A_i \mid A_i[-1])$. *Since the mutation map* $\texttt{mut}$ *already raises the Morse indices by one we determine* $\gamma_i - 1$ *to be the least number of suspensions. Let*

$$\Gamma_i \geq \max_{i=1,\ldots,M}\{\gamma_i\} - 1.$$

We obtain M *Morse meanders of the form*[2]*:*

$$\begin{aligned}\texttt{sus}_{\Gamma_i}(\texttt{mut}(\mathcal{M}(A_i \mid A_i[-1]))) &= \texttt{sus}_{\Gamma_i}(\texttt{shift}_{[1]}(\mathcal{M}(1, A_i \mid A_i, 1)) \\ &= \texttt{shift}_{[1]}(\mathcal{M}(\Gamma_i(1, A_i) \mid \Gamma_i(A_i, 1))).\end{aligned}$$

We may concatenate the recovered left one-shifted elements $\mathcal{M}(A_i \mid A_i[-1])$ *by identifying the final vertex of*

$$\texttt{shift}_{[1]}(\mathcal{M}(\Gamma_i(1, A_i) \mid \Gamma_i(A_i, 1)), \; i = 1, \ldots, M-1$$

with the initial vertex of

$$\texttt{shift}_{[1]}(\mathcal{M}(\Gamma_{i+1}(A_{i+1}, 1) \mid \Gamma_{i+1}(A_{i+1}, 1))), \; i = 1, \ldots, M-1.$$

By right concatenation $\circ_{\mathrm{R}}$*, we get following Morse meanders:*

$$\begin{aligned}&\texttt{shift}_{[1]}(\mathcal{M}(\,\Gamma_1(1, A_1), \ldots, \Gamma_M(1, A_M) \mid \Gamma_1(A_1, 1), \ldots, \Gamma_M(A_M, 1))) = \\ &\mathcal{M}^{coll}(2\Gamma_1(2, 2A_1), \ldots, 2\Gamma_M(2, 2A_M), 1 \mid 1, 2\Gamma_1(2A_1, 2), \ldots, 2\Gamma_M(2A_M, 2))\end{aligned}$$

B) *We concatenate* M *left one-shifted elements* $\mathcal{M}(A_i \mid A_i)$, $i = 1, \ldots, M$ *by identifying the last vertex of*

$$\mathcal{M}(A_i \mid A_i), i = 1, \ldots, M-1$$

with the first vertex of $\mathcal{M}(A_{i+1} \mid A_{i+1})$, $i = 1, \ldots, M-1$. *After concatenating, we obtain* $\mathcal{M}(A_1, \ldots, A_M \mid (A_1, \ldots, A_M)[-1])$. *In order to recover negative Morse indices, we use Proposition 4.2 and subtract one from* γ_k*, which is again by the property of the mutation map* $\texttt{mut}$. *Let*

$$\Gamma_k \geq \gamma_k - 1 = \max_{i=1,\ldots,M}\{\gamma_i\} - 1.$$

We end up with Morse meanders of the form:

$$\begin{aligned}&\texttt{sus}_{\Gamma_k}(\texttt{mut}(\mathcal{M}(A_1, \ldots, A_M \mid (A_1, \ldots, A_M)[-1]))) \\ &= \texttt{sus}_{\Gamma_k}(\texttt{shift}_{[1]}(\mathcal{M}(1, A_1, \ldots, A_M \mid A_1, \ldots, A_M, 1))) \\ &= \texttt{sus}_{\Gamma_k}(\mathcal{M}^{coll}(2, 2A_1, \ldots, 2A_M, 1 \mid 1, 2A_1, \ldots, 2A_M, 2)) \\ &= \mathcal{M}^{coll}(2\Gamma_k(2, 2A_1, \ldots, 2A_M), 1 \mid 1, 2\Gamma_k(2A_1, \ldots, 2A_M, 2)).\end{aligned}$$

2 Recall from Chapter 2, that we have to multiply the entries standing for the number of nodes inside of collapsed meanders $\mathcal{M}^{coll}$ when switching from meanders $\mathcal{M}$ by two.

C) *We consider concatenations of*

$$\mathtt{mut}(\mathcal{M}(A_i \mid A_i[-1])),\ i = K,\dots,N, \quad K, N \in \mathbb{N}_{>0}$$

with Morse meanders obtained in A) or B) and then suspend sufficiently many times in order to recover negative Morse indices of

$$\mathtt{mut}(\mathcal{M}(A_i \mid A_i[-1])) = \mathcal{M}^{coll}(2, 2A_i, 1 \mid 1, 2A_i, 2).$$

In order to determine the least number of suspensions, we again use Proposition 4.2. We subtract one from the determined number Γ_k, $k \in \{K,\dots,N\}$ since we have applied the map mut *on $\mathcal{M}(A_i \mid A_i[-1])$.*

D) *We concatenate Morse meanders of the form described in A), B), or A),C) or B), C) or A), B), C).*

E) *We suspend Morse meanders in D) arbitrary many times.*

In Figure 4.5, we give examples of Morse meanders of type II for different subclasses of the above construction listed in i), $i = A, B, C, D$. We have omitted examples for Morse meanders of type II constructed in E).

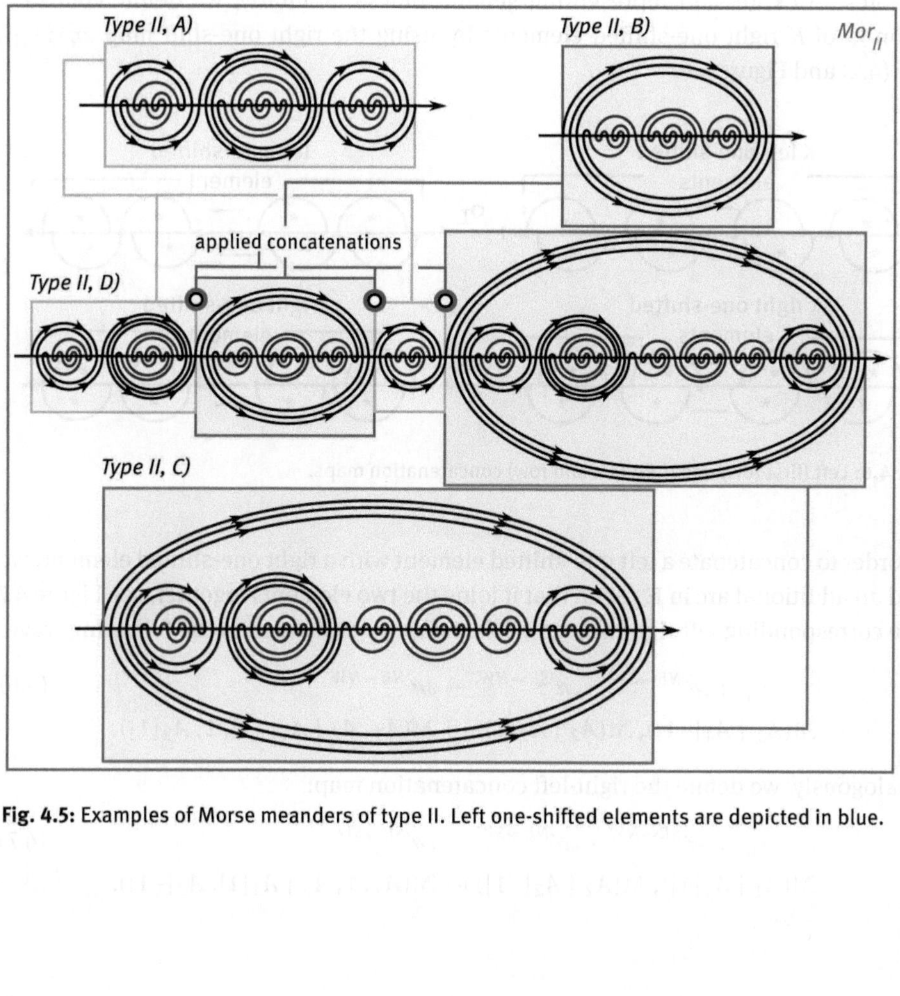

Fig. 4.5: Examples of Morse meanders of type II. Left one-shifted elements are depicted in blue.

4.3 Morse Meanders of Type III and IV

Let us define maps which allow us to concatenate K elements with each other. It is clear that two elements of the same type, say

$$\mathtt{shift}_{[1]}(\mathcal{M}(A \mid A)) \text{ and } \mathtt{shift}_{[1]}(\mathcal{M}(A' \mid A'))$$

or $\mathtt{shift}_{[-1]}(\mathcal{M}(A \mid A))$ and $\mathtt{shift}_{[-1]}(\mathcal{M}(A' \mid A'))$ may be concatenated by one-shifting $\mathcal{M}(A, A' \mid A, A')$ to the right, respectively to the left, see Figure 4.6. Formally, we define concatenation of K left, respectively K right one-shifted elements in the following way:

$$\begin{aligned} \circ_{\mathrm{L}} : (\mathscr{M}^{NE\rightsquigarrow SW})^{\times K} &\to \mathscr{M}^{NE\rightsquigarrow SW} \\ \mathcal{M}(A_1 \mid A_1[-1]), \ldots, \mathcal{M}(A_K \mid A_K[-1]) &\mapsto \mathcal{M}(A_1, \ldots, A_K \mid (A_1, \ldots A_K)[-1]) \\ &=: \mathcal{M}(A_1 \mid A_1[-1]) \circ_{\mathrm{L}} \ldots \circ_{\mathrm{L}} \mathcal{M}(A_K \mid A_K[-1]), \end{aligned} \tag{4.5}$$

where

$$A_i, \; i = 1, \ldots, K,$$

denotes an expression representing general blocks. Similarly, we define concatenation $\circ_{\mathrm{R}}$ of K right one-shifted elements by using the right one-shift map $\mathtt{shift}_{[1]}$, see (4.2) and Figure 4.6.

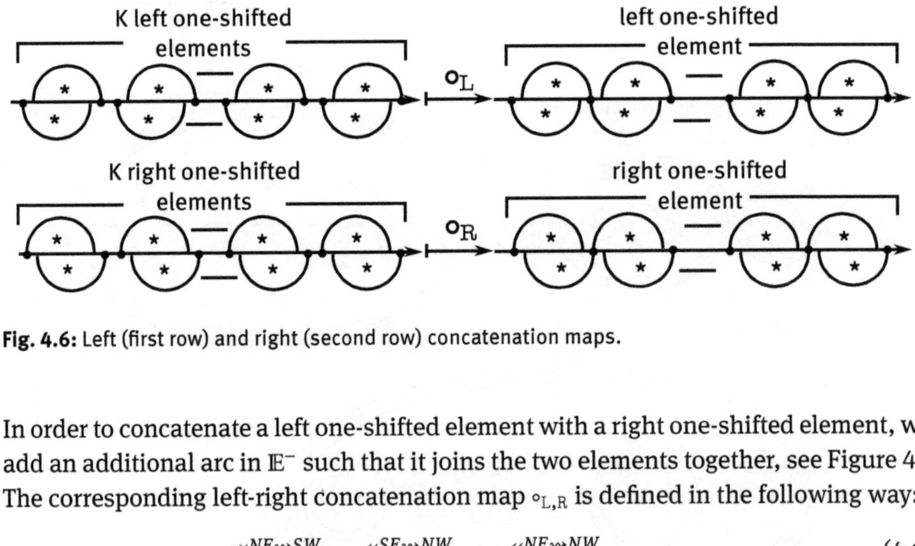

Fig. 4.6: Left (first row) and right (second row) concatenation maps.

In order to concatenate a left one-shifted element with a right one-shifted element, we add an additional arc in $\mathbb{E}^-$ such that it joins the two elements together, see Figure 4.7. The corresponding left-right concatenation map $\circ_{\mathrm{L,R}}$ is defined in the following way:

$$\begin{aligned} \circ_{\mathrm{L,R}} : \mathscr{M}^{NE\rightsquigarrow SW} \times \mathscr{M}^{SE\rightsquigarrow NW} &\to \mathscr{M}^{NE\rightsquigarrow NW} \\ \mathcal{M}(A_1 \mid A_1[-1]), \mathcal{M}(A_2 \mid A_2[1]) &\mapsto \mathcal{M}(A_1, A_2 \mid A_1[-1], 1, A_2[1]). \end{aligned} \tag{4.6}$$

Analogously, we define the right-left concatenation map:

$$\begin{aligned} \circ_{\mathrm{R,L}} : \mathscr{M}^{SE\rightsquigarrow NW} \times \mathscr{M}^{NE\rightsquigarrow SW} &\to \mathscr{M}^{SE\rightsquigarrow SW} \\ \mathcal{M}(A_1 \mid A_1[1]), \mathcal{M}(A_2 \mid A_2[-1]) &\mapsto \mathcal{M}(A_1, 1, A_2 \mid A_1[1], A_2[-1]). \end{aligned} \tag{4.7}$$

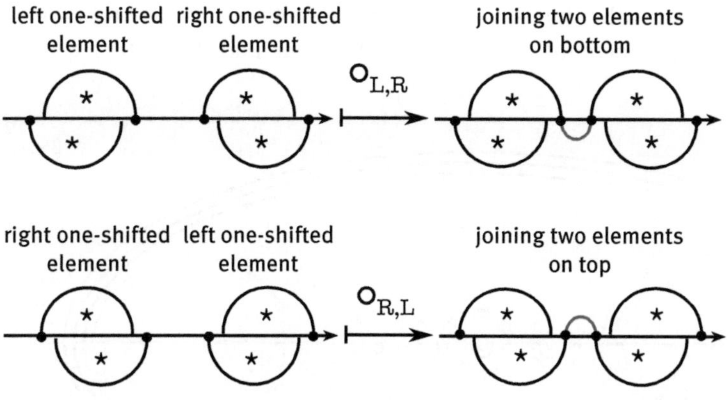

Fig. 4.7: Left-right (first row) and right-left (second row) concatenation maps.

In particular by definition of the maps $\circ_{L,R}$, $\circ_{R,L}$, $\mathtt{mut}$ and $\mathtt{shift}_{[1]}$, we obtain the following correspondencies:

$$\ldots \circ_{R,L} \mathcal{M}(A \mid A[-1]) \circ_{L,R} \ldots = \ldots \circ_R \mathtt{mut}(\mathcal{M}(A \mid A[-1])) \circ_R \ldots \tag{4.8}$$

$$= \ldots \circ_R \mathtt{shift}_{[1]}(\mathcal{M}(1, A \mid A, 1)) \circ_R \ldots, \tag{4.9}$$

where "$\ldots$" stands for extending the sequence by elements from south-west to north-east on both sides. With the aid of concatenations $\circ_{L,R}$ and $\circ_{L,R}$, we introduce the following construction of Morse meanders of *type III*:

Construction 9 (Morse Meanders of Type III). *Let*

$$A_i \neq (1, \ldots, 1),$$

also $A_i \neq 1$, be a bracket expression representing a general block. By using $\circ_{L,R}$, $\circ_{R,L}$, we concatenate left one-shifted elements

$$\mathcal{M}(A_i \mid A_i[-1])$$

with right one-shifted elements and then suspend sufficiently many times as described in Proposition 4.2. Note, that in order to apply the map $\mathtt{sus}_{(\cdot)}$ we need to append one arc in $\mathbb{E}^+$ if the first element is a left one-shifted one and one arc in $\mathbb{E}^-$ if the last element is a left one-shifted one. In each case, the least number of suspensions decreases by one.

Note that the above construction may be introduced without the right-left and left-right concatenation maps, $\circ_{R,L}$ and $\circ_{R,L}$. Instead, we could have required to concatenate right-one shifted elements with *mutated* left one-shifted elements

$$\mathtt{mut}(\mathcal{M}(A \mid A[-1]))$$

by using the right concatenation map $\circ_R$. This holds by the correspondence listed in (4.8) and the fact that the suspension map is applicable to the elements of the form

$\mathtt{mut}(\mathcal{M}(A \mid A[-1]))$. On the other hand, we could even forget about the mutation map $\mathtt{mut}$ by just using the right one-shift $\mathtt{shift}_{[1]}$ instead, which holds by (4.9).

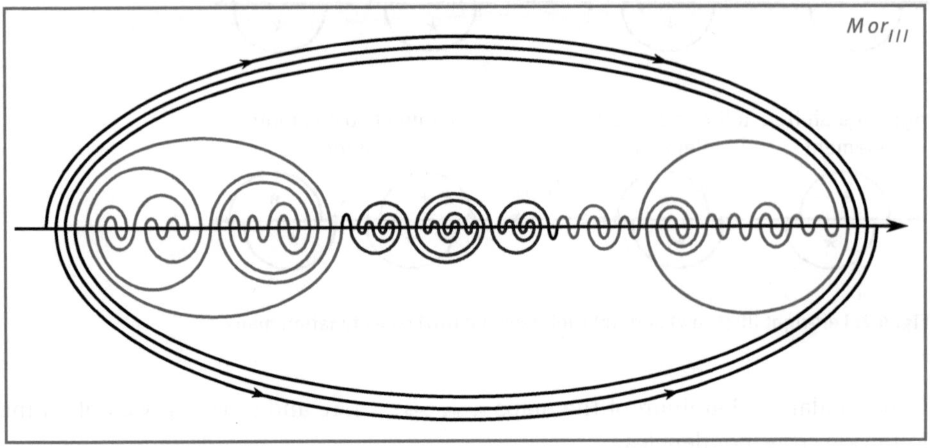

Fig. 4.8: Example of a Morse meander of type III.

Construction 10 (Morse Meanders of Type IV). *Let $\mathcal{M}_i \in Mor_I$, $i = 1, \dots, N$, be Morse meanders of type I, $\mathcal{M}_j \in Mor_{II}$, $j = 1, \dots, M$, be Morse meanders of type II and $\mathcal{M}_l \in Mor_{III}$, $l = 1, \dots, K$, be Morse meanders of type III. Morse meanders of type IV are concatenations of mixed types, i.e., no concatenations of only one type are taken into account. We may suspend the resulting Morse meander after concatenation.*

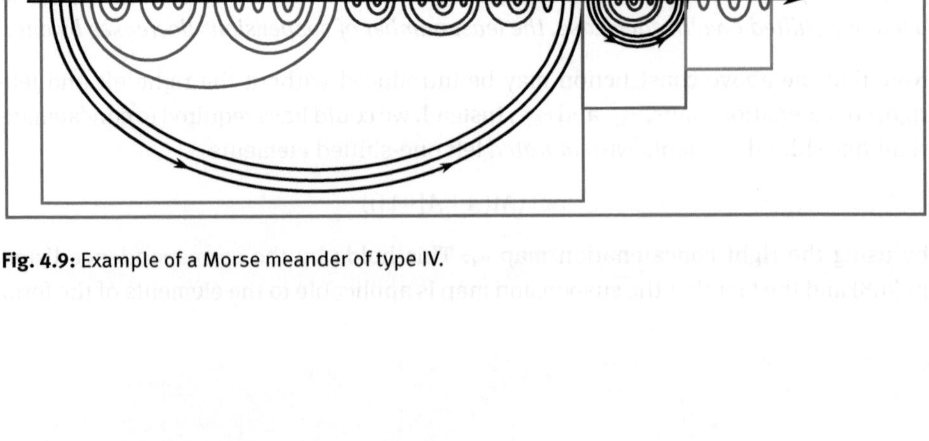

Fig. 4.9: Example of a Morse meander of type IV.

We summarize the above Constructions 7-10 of Morse meanders of type I–IV. By generalizing Constructions 7-10, it is possible to introduce Morse meanders of type χ. For this reason, let us consider the following table of Morse meanders of the four types expressed in terms of the right one-shift.

Table 4.1: Short overview of the previous constructions for *Mor I – Mor IV*

Mor I	$\mathtt{shift}_{[1]}(\mathcal{M}(A \mid A))$
Mor II	*Let* $A_l \neq (1,\dots,1), A_1 \neq 1$ *then consider subsets Mor II, A)–E)*
Mor II, A)	$\mathtt{shift}_{[1]}(\mathcal{M}(\Gamma_1(1,A_1),\dots,\Gamma_n(1,A_n) \mid \Gamma_1(A_1,1),\dots,\Gamma_n(A_n,1))$
Mor II, B)	$\mathtt{shift}_{[1]}(\mathcal{M}(\Gamma_i(1,A_1,\dots,A_n) \mid \Gamma_i(A_1,\dots,A_n,1)),\ \Gamma_i := \max_{j=1,\dots,n} \Gamma_j$
Mor II, C-E)	*arbitrary concatenations and suspensions of Mor II, A), B)*
Mor III	$\mathtt{shift}_{[1]}(\mathcal{M}(\Gamma_j(\dots,1,A_j,\dots) \mid \Gamma_j(\dots,A_j,1,\dots))$
Mor IV	*concatenations of two different types (I–III) of Morse meanders*

In order to generalize the above table – more precisely the construction of *Mor III* – we may consider concatenations of

$$\mathtt{mut}(\mathtt{shift}_{[-1]}(\mathcal{M}(A_j \mid A_j))) = \mathtt{shift}_{[1]}(\mathcal{M}(1,A_j \mid A_j,1))$$

with Morse meanders of type IV and then suspend sufficiently many times. By concatenating Morse meanders of at least two different types $i,j \in \{I,\dots,IV\}$ excluding combinations occurring in the set of Morse meanders of type IV, we obtain the class *Mor V*. By repeating these steps, we obtain a large number of different Morse meanders for sufficiently large, fixed and odd number of intersection points with the straight line. We call the whole class of Morse meanders constructed by combining left and right one-shifts ($\mathtt{shift}_{[-1]}$ and $\mathtt{shift}_{[1]}$), mutation map $\mathtt{mut}$, concatenations $\circ_{L,R}$, $\circ_{R,L}$, $\circ_R$, $\circ_L$ and suspension map $\mathtt{sus}_{(\cdot)}$ applied – as described for classes of Morse meanders of different types – to closed meanders with same compositions for upper and lower blocks *class of Morse meanders of type* χ, and denote it by Mor_χ.

4.4 Sets of Morse Meanders of Type I, II, III, and IV are Disjoint and Consist of Pitchforkable Elements

Before formulating and proving Theorem 4.2, we recall the following situations for equilibria v from the Figure 4.10 shown below. The numbers next to equilibria denote the corresponding Morse indices. Recall from Chapter 4, that the permutation of the right repectively left identity elements, denoted by

$$\mathtt{shift}_{[1]}(\mathcal{M}(1 \mid 1)) \text{ and } \mathtt{shift}_{[-1]}(\mathcal{M}(1 \mid 1))$$

is given by $id \in S_3$. By the *contraction* of right respectively left identity elements on three vertices, we mean the reduction of the correspnding elements to one single vertex, as depicted in Figure 4.10, see the last row.

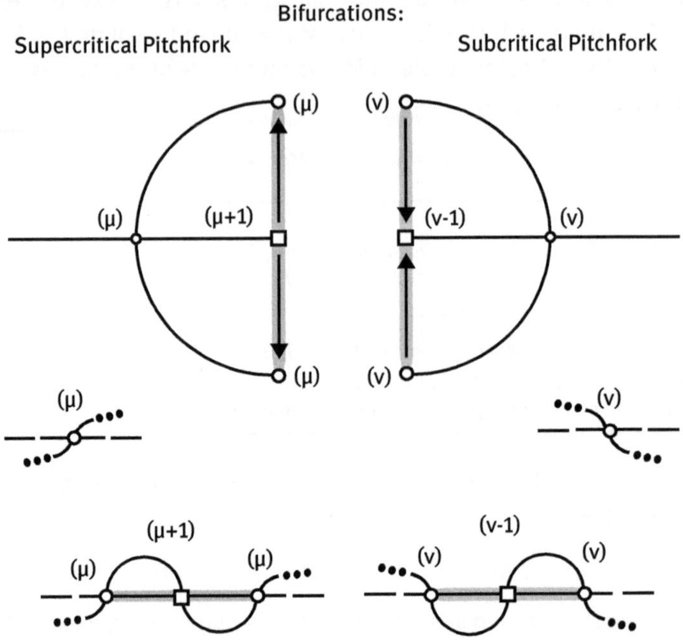

Fig. 4.10: Supercritical and subcritical pitchfork bifurcations and the corresponding right and left identity elements on three vertices depicted in the middle below and the corresponding contraction to one single vertex which is next to the right/left identity element from the left/right .

Definition 4.4. *Let $n > 1$ be odd. A Morse meander $\mathcal{M}$ on n vertices is called* pitchforkable, *if we may successively perform a finite number of contractions of right respectively left identity elements on three vertices, in order to reduce $\mathcal{M}$ to one single vertex.*

The above definition of pitchforkable Morse meanders is in correspondence with the requirements imposed to the curves appearing in the sequence of curves $\gamma_0, \ldots, \gamma_K$ in Definition 3.6 of [15], since the right identity element on three vertices is equivalent to γ^+ and the left identitity element on three vertices is equivalent to γ^-.

Let $Mor_{I,II,III,IV}$ denote the set of Morse meanders of type I, II, III, IV. We prove the following theorem.

Theorem 4.2 (Pitchforkability and Partition). *Let $Mor_j, j = I, II, III, IV$ denote the set of Morse meanders of type j. Then, for $j \neq k \in \{I, II, III, IV\}$ it holds*

i) *Morse meanders of type j are pitchforkable,*
ii) *$Mor_j \cap Mor_k = \emptyset$. In other words the set $Mor_{I,II,III,IV}$ is partitioned into the subsets of Morse meanders of type I, II, III, and IV, i.e,*

$$Mor_{I,II,III,IV} = Mor_I \sqcup Mor_{II} \sqcup Mor_{III} \sqcup Mor_{IV}.$$

Proof. i) Morse meanders of type I given by

$$\texttt{shift}(\mathcal{M}(A \mid A)) = \mathcal{M}(A \mid A[1])$$

are pitchforkable, which follows by induction on n denoting the number of arcs contained in the block represented by A.

Induction start: $n = 1$. The permutation of Morse meander $\mathcal{M}(1 \mid 1[1])$, which is the right identity element on three vertices, is given by id_{S_3}. Contracting the right identity element to one vertex, which represents an equilibrium, corresponds to a supercritical pitchfork bifurcation, since two stable equilibria, vertices with Morse index zero, become one single equilibrium. Therefore, the right identity element $\mathcal{M}(1 \mid 1[1])$ is pitchforkable.

Induction step: $n \curvearrowright n + 1$. Let A' represent a block which has $n + 1$ arcs. We select a *shortest arc*[3], i.e., an arc which joins consecutive vertices on the straight line, in the block represented by A'. After the right one-shift applied to $\mathcal{M}(A' \mid A')$, the right one-shift of the corresponding lower and upper shortest arcs yields a right identity element $\mathcal{M}(1 \mid 1[1])$ inside of $\mathcal{M}(A' \mid A'[1])$. By contracting the right identity element to one vertex inside of $\mathcal{M}(A' \mid A'[1])$, which corresponds to a supercritical pitchfork bifurcation, we obtain $\mathcal{M}(A \mid A[1])$ which is pitchforkable by the induction assertion. Hence, $\mathcal{M}(A' \mid A')$ is pitchforkable.

In order to prove that Morse meanders of type II are pitchforkable, we note that Morse meanders of type II are constructed by using concatenations and suspensions of elements of the form $\texttt{mut}(\mathcal{M}(A_k \mid A_k[-1]))$ with $A_k \neq (1, \ldots, 1)$ and $A_k \neq 1$ representing a block. The left identity element $\mathcal{M}(1 \mid 1[-1])$ on three vertices is obtained from an equilibrium which undergoes the subcritical pitchfork bifurcation. By the identity given in (4.4) and since we have already proved that $\texttt{shift}_{[1]}(\mathcal{M}(A \mid A))$ are pitchforkable, we know for elements of the form $\texttt{mut}(\mathcal{M}(A_k \mid A_k[-1]))$ that after finitely many contractions performed inside of left one-shifted elements $\mathcal{M}(A_k \mid A_k[-1])$ – by starting with $\mathcal{M}(1 \mid 1[-1])$ inserted in $\mathcal{M}(A_k \mid A_k[-1])$ – we obtain one single vertex. Therefore after finitely many steps, elements of the form $\texttt{mut}(\mathcal{M}(A_k \mid A_k[-1]))$ become the right identity element $\mathcal{M}(1 \mid 1[1])$ which is pitchforkable by induction start in the beginning of the proof of part i) for the Morse meanders of type I. Therefore, Morse meanders of type II are pitchforkable.

Morse meanders of type III are obtained by suspensions of concatenated

$$\texttt{mut}(\mathcal{M}(A_k \mid A_k[-1]))$$

with Morse meanders of type I. Since both types of elements are pitchforkable, Morse meanders of type III are pitchforkable. Analogously for Morse meanders of type IV, it holds by construction that they are pitchforkable, since they were obtained by concatenating at least two different types out of the types I, II and III for which we have already seen that they are pitchforkable. This finishes part i).

3 This was called short arc in [29].

ii) We first show that $Mor_I \cap Mor_{II} = \emptyset$. It suffices to show that an element of the set Mor_{II} is not in the set Mor_I. The elements of Mor_{II} contain suspensions of the elements of the following form $\mathtt{mut}(\mathcal{M}(A \mid A[-1]))$, where A is a bracket expression either representing one block or several blocks. We consider the following identities, in order to determine the necessary and sufficient condition for writing down $\mathtt{mut}(\mathcal{M}(A \mid A[-1]))$ in terms of the right one-shift: It holds

$$\begin{aligned} \mathtt{mut}(\mathcal{M}(A \mid A[-1])) &= \mathtt{shift}_{[1]}(\mathcal{M}(1, A \mid A, 1)) &(4.10)\\ &= \mathtt{shift}_{[1]}(\mathcal{M}(A' \mid A')) &(4.11)\end{aligned}$$

if and only if $A' = (1, A)$ and $A' = (A, 1)$, i.e., $A = 1$ or $A = (1, \dots, 1)$. Since we have excluded $A = 1,\ A = (1, \dots, 1)$ in Construction 8 of Morse meanders of type II, it holds $Mor_I \cap Mor_{II} = \emptyset$. In order to prove, that the set of Morse meanders of type III is disjoint with the set of Morse meanders of type I or II, i.e.,

$$Mor_{III} \cap Mor_i = \emptyset$$

for $i = I, II$, it suffices to consider three cases: Let $B \neq (1, \dots, 1),\ B \neq 1$ as required for compositions of elements which belong to Mor_{III} and let the following elements be in the set Mor_{III}

i) $\mathtt{sus}_{\Gamma_B}(\mathcal{M}(A \mid A[1]) \circ_{R,L} \mathcal{M}(B \mid B[-1]) \circ_{L,R} \mathcal{M}(C \mid C[1])) \in Mor_{III}$
ii) $\mathtt{sus}_{\Gamma_B}\big(\mathcal{M}(A \mid A[1]) \circ_{R} \mathtt{mut}(\mathcal{M}(B \mid B[-1]))\big) \in Mor_{III}$
iii) $\mathtt{sus}_{\Gamma_B}\big(\mathtt{mut}(\mathcal{M}(B \mid B[-1])) \circ_{R} \mathcal{M}(A \mid A[1])\big) \in Mor_{III}$,

where Γ_B is the least number of suspensions required for raising the Morse indices of left one-shifted element $\mathcal{M}(B \mid B[-1])$ in such a way that all Morse indices of the resulting meander in i), ii), iii) are positive. First note, that in general each element of Mor_{III} arise from concatenations of finitely many components as listed in the above three cases. We either have left one-shifted elements between right one-shifted elements (case i)), or we have left one-shifted elements at the end (case ii)) or at the beginning (case iii)) or both at the end and the beginning which results from a combination of ii) and iii). Alternating sequences of left and right one-shifted elements may be omitted, since the argumentation is analogous to the three cases listed above. Also note, that the third case is analogous to the second one. For case i), ii) we show that

$$\mathtt{sus}_{\Gamma_B}(\mathcal{M}(A \mid A[1])) \circ_{R,L} \mathcal{M}(B \mid B[-1]) \circ_{L,R} \mathcal{M}(C \mid C[1])) \notin Mor_x$$

and

$$\mathtt{sus}_{\Gamma_B}\big(\mathcal{M}(A \mid A[1]) \circ_{R} \mathtt{mut}(\mathcal{M}(B \mid B[-1]))\big) \notin Mor_x, \quad x = I, II.$$

For this it suffices to consider the following,

$$\begin{aligned} &\mathtt{sus}_{\Gamma_B}(\mathcal{M}(A \mid A[1])) \circ_{R,L} \mathcal{M}(B \mid B[-1]) \circ_{L,R} \mathcal{M}(C \mid C[1]))\\ &= \mathtt{shift}_{[1]}\big(\mathcal{M}(\Gamma_B(A, 1, B, C) \mid \Gamma_B(A, B, 1, C))\big) \notin Mor_I,\end{aligned}$$

since, similar to the argument for

$$Mor_I \cap Mor_{II} = \emptyset,$$

we have $B \neq (1, \dots, 1)$, $B \neq 1$ and therefore $(A, 1, B, C) \neq (A, B, 1, C)$ which is not allowed for elements from the set Mor_I, since they were constructed by applying the right one-shift map to closed meanders with the *same* compositions of lower and upper blocks. Moreover, we have that

$$\begin{aligned}
&\mathtt{sus}_{\Gamma_B}(\mathcal{M}(A \mid A[1])) \circ_{\mathrm{R,L}} \mathcal{M}(B \mid B[-1]) \circ_{\mathrm{L,R}} \mathcal{M}(C \mid C[1])) \\
&= \mathtt{sus}_{\Gamma_B}\Big(\mathtt{shift}_{[1]}(\mathcal{M}(A, 1, B, C \mid A, B, 1, C))\Big), \\
&= \mathtt{sus}_{\Gamma_B}\Big(\mathtt{shift}_{[1]}(\mathcal{M}(1, A' \mid A', 1))\Big) \\
&= \mathtt{sus}_{\Gamma_B}\Big(\mathtt{mut}(\mathtt{shift}_{[-1]}(\mathcal{M}(A' \mid A'))\Big)
\end{aligned}$$

if and only if

$$(A, 1, B, C) = (1, A') \text{ and } (A, B, 1, C) = (A', 1)$$

which is fulfilled for

$$A = \underbrace{(1, \dots, 1)}_{k} \text{ and } C = \underbrace{(1, \dots, 1)}_{k}.$$

Again

$$A \neq (1, \dots, 1),\ A \neq 1 \text{ and } C \neq (1, \dots, 1),\ C \neq 1$$

were required in the construction of Morse meanders belonging to the set Mor_{II}. Therefore,

$$\mathtt{sus}_{\Gamma_B}(\mathcal{M}(A \mid A[1])) \circ_{\mathrm{R,L}} \mathcal{M}(B \mid B[-1]) \circ_{\mathrm{L,R}} \mathcal{M}(C \mid C[1])) \notin Mor_{II}.$$

The arguments for the last two cases ii) and iii) are analogous to the first case i), since

$$\dots \circ_{\mathrm{R,L}} \mathcal{M}(B \mid B[-1]) \circ_{\mathrm{L,R}} \dots = \dots \circ_{\mathrm{R}} \mathtt{mut}(\mathcal{M}(B \mid B[-1])) \circ_{\mathrm{R}} \dots,$$

where " . . . " stands for possible right concatenations with right one-shifted elements, see (4.8). By the above calculations, it follows that the set of Morse meanders of type *III* is disjoint with the set of Morse meanders of type *I* or *II*, i.e.,

$$Mor_{III} \cap Mor_x = \emptyset \text{ for } x = I, II.$$

Let $\mathcal{M}_x \in Mor_x$, $x = I, II, III$. We recall that Morse meanders of type *IV* are obtained by concatenating elements of at least two different types, i.e., up to the order Morse meanders of type *IV* are of the following form

a) $\dots \mathcal{M}_I \circ_{\mathrm{R}} \mathcal{M}_{II} \dots \in Mor_{IV}$
b) $\dots \mathcal{M}_{II} \circ_{\mathrm{R}} \mathcal{M}_{III} \dots \in Mor_{IV}$
c) $\dots \mathcal{M}_I \circ_{\mathrm{R}} \mathcal{M}_{III} \dots \in Mor_{IV}$
d) $\dots \mathcal{M}_I \circ_{\mathrm{R}} \mathcal{M}_{II} \circ_{\mathrm{R}} \mathcal{M}_{III} \dots \in Mor_{IV}$.

Since

$$\mathcal{M}_x \notin Mor_y \text{ for } x \neq y,$$

which is by the preceding arguments of the proof and because the right concatenation $\circ_R$ is joining two different types of elements together, it follows that the elements listed in a), b), c), d) are not in the set Mor_z for $z = I, II, III$. Therefore, it holds that

$$Mor_j \cap Mor_{IV} = \emptyset \text{ for } j = I, II, III.$$

□

Remarks 4.2.

i) *Theorem 4.2 implies that the permutations σ_j of a Morse meander of type*

$$j, \quad j \in \{I, II, III, IV\}$$

are in the class of admissible permutations Π as defined in [15] . Therefore, a combination of Theorem 4.2 and Theorem 3.3 by G. Fusco and C. Rocha implies that there is a smooth function

$$f : [0, 1] \times \mathbb{R}^2 \to \mathbb{R}$$

such that σ_j is a Sturm permutation, i.e., corresponds to the shooting curve γ_f.

ii) *We do not construct all pitchforkable Morse meanders obtained by G. Fusco and C. Rocha as recalled from [15], p. 52. For example, consider: Morse meander given by*

$$\mathtt{shift}_{[1]}(\mathcal{M}(3 \mid 1, 2))$$

in Figure 4.11. However in Chapter 5, by explicit formulation of the classes of Morse meanders of type I, II, III, and IV we do obtain results which answer isomorphism questions with respect to the underlying connection graphs obtained from the underlying Morse meanders of types I, II, III, IV.

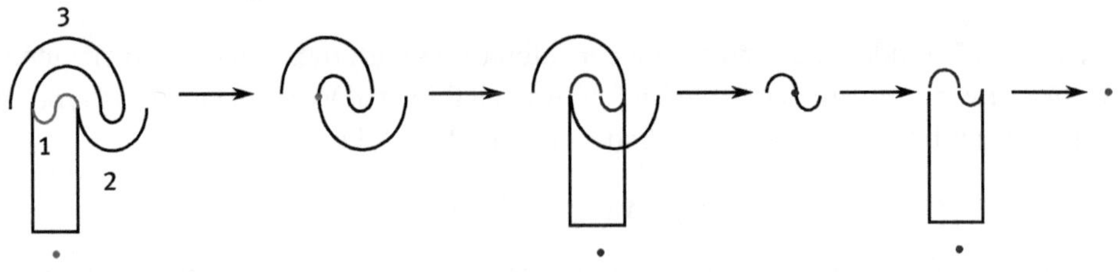

Fig. 4.11: An example of a pitchforkable Morse meander which is not contained in the class of Morse meanders of type I, II, III or IV; dashed open square around left (in red) or right (in blue) identity elements indicate contraction to a vertex depicted below.

4.5 Detecting Morse Meanders – Discussion

By applying the right and left one-shifts to closed meanders with same compositions for lower and upper blocks, we do *not* construct all Morse meanders, which may be deduced from the last Morse meander $\mathcal{M}^{coll}(6, 1 \mid 1, 2, 4)$ in Figure 3.5. In this section we therefore consider the next interesting case of meanders

$$\mathcal{S}^{coll}(2(A + B), 1 \mid 1, 2A, 2B) = \mathtt{shift}_{[1]}(A + B \mid A, B),$$

where A, B are positive integers representing rainbow blocks. The number of connected components of $\mathtt{shift}_{[1]}(\mathcal{S}(A + B \mid A, B))$ is given by $\mathtt{gcd}(A + 1, B - 1)$ which is by the following calculation:

$$\begin{aligned} \mathtt{Z}(\mathcal{S}(1, A, B \mid A + B + 1)) &= \mathtt{gcd}(A + 1, B + A) \\ &= \mathtt{gcd}(A + 1, B + A - (A + 1)) \\ &= \mathtt{gcd}(A + 1, B - 1). \end{aligned}$$

In the above calculation we have used that the number of connected components of $\mathtt{shift}_{[1]}(\mathcal{S}(A + B \mid A, B))$ equals the number of connected components of

$$\mathcal{S}(1, A, B \mid A + B + 1) = \mathcal{S}(1, A, B)$$

In Figure 4.12, we depict the number of connected components of

$$\mathtt{shift}_{[1]}(\mathcal{S}(A + B \mid A, B)) \text{ for } A \neq B.$$

The problem of determining connected seaweed meanders may be handled as follows: By applying the inverse collapse map

$$\mathtt{coll}^{-1} \text{ to } \mathcal{S}(A + B \mid A, B[1]) = (\mathtt{shift}_{[1]}(\mathcal{S}(A + B \mid A, B))$$

we are able to determine the number of connected components of

$$\mathtt{coll}^{-1}(\mathtt{shift}_{[1]}(\mathcal{S}(A + B \mid A, B)))$$

by using algorithms presented in [18]. In addition, not all connected meanders

$$\mathtt{shift}_{[1]}(\mathcal{S}(A + B \mid A, B))$$

are Morse. Consider the families given by $\mathcal{S}(A + 2 \mid (A, 2)[1]),\ A > 1$ and

$$\mathcal{S}(2A + 3 \mid (A, A + 3)[1])$$

with maximal negative Morse index $-(A - 1)$. In order to recover Morse meanders from these families we need to suspend at least $(A - 2)$-many times after applying the map $\mathtt{mut}$ to $\mathcal{S}(2A + 3 \mid (A, A + 3)[1])$, which already raises all Morse indices by one.

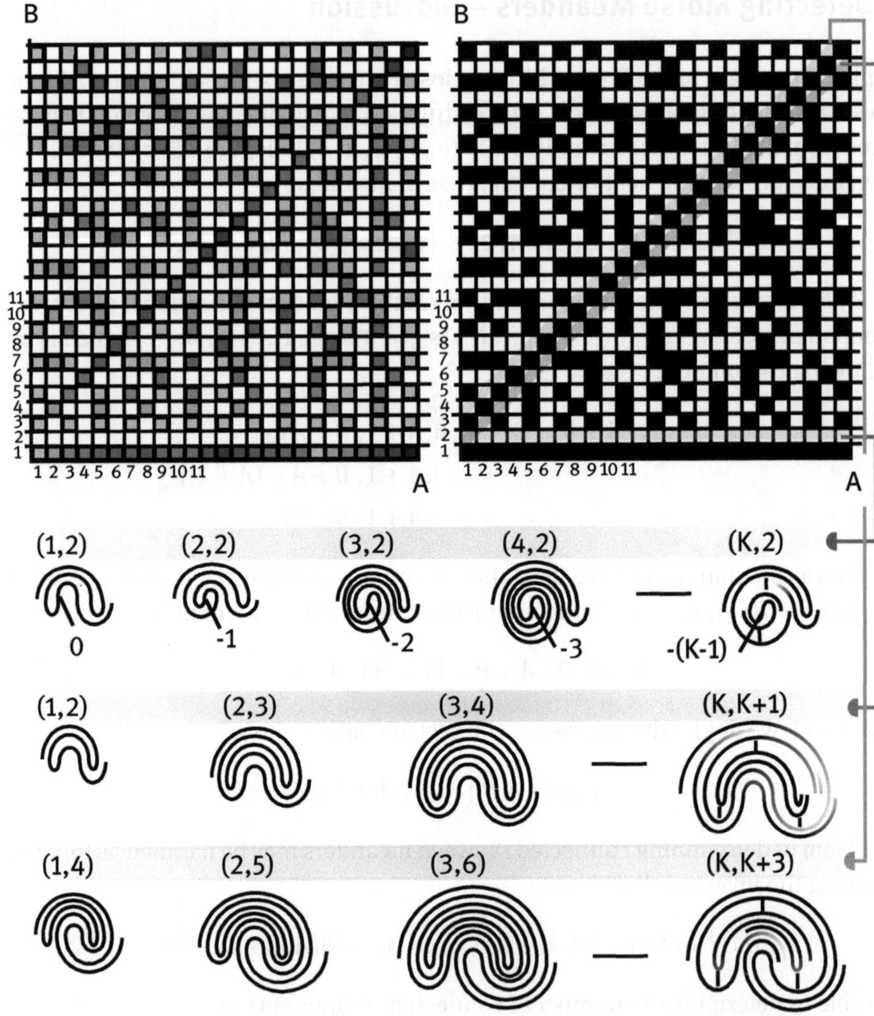

Fig. 4.12: Left: Colored diagram for the number of connected components of right one-shifted meanders $\mathcal{S}(A + B \mid (A, B)[1])$, white squares signify connected open meanders; right: the same diagram in black and white; below: selected examples on diagonals $(A, A + 1)$, $(A, A + 3)$ and the row $(A, 2)$ containing only connected meanders which are not necessarily Morse (compare with negative Morse indices for $(A, 2)$, $A > 1$).

The following family

$$\mathcal{S}(2A + 1 \mid (A, A + 1)[1])$$

of connected meanders corresponding to the tuples $(A, A + 1)$ are Morse.

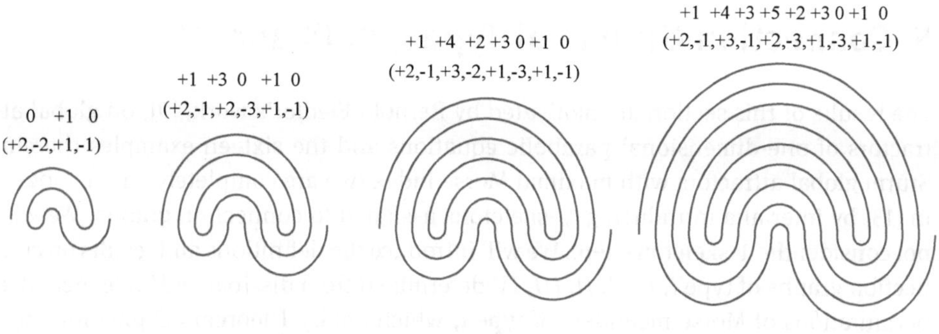

Fig. 4.13: Morse meanders corresponding to the diagonal $f(A) = A + 1$; the sequence of right turns (green) and left turns (red) is written above the meanders.

Explicitly, the sequences of left and right turns corresponding to seaweed meanders

$$\mathcal{S}(2A + 1 \mid (A, A + 1)[1])$$

are given by:

$$(1, 2) : (2, -2, 1, -1)$$

$$(2K, 2K + 1) :$$

$$(2, -1, \underbrace{3, -1, \ldots 3, -1}_{2K-2}, 2, \underbrace{3, -1, \ldots 3, -1}_{2K-2}, 3, -1, 1), \quad K > 1$$

$$(2K + 1, 2K + 2) :$$

$$(2, -1, \underbrace{3, -1, \ldots 3, -1}_{2K-2}, 3, -2, \underbrace{1, -3, \ldots 1, -3}_{2K-2}, 1, -3, 1, -1), K > 1.$$

Since the alternating sums of each pair of numbers of left and right turns in the above sequences are positive, all Morse indices are positive. The above example serves for an illustration of the complexity of determining a general condition on $\mathcal{S}(A+B \mid (A, B)[1])$ for being Morse meanders. They produce examples of Morse meanders which are non-pitchforkable, compare with an example in [9] given by the permutation

$$\pi = (1\ 6\ 7\ 10\ 3\ 4\ 9\ 8\ 5\ 2\ 11).$$

We therefore call the undetermined class of Morse meanders, which arise by applying the one-shift maps to meanders with unequal compositions for lower and upper blocks, *Morse meanders of type* $\overline{\chi}$. Conversely, we call Morse meanders of type I, II, III, IV, ... of *type* χ. Indeed, it is a challenging question how to classify combinatorially Morse meanders of type $\overline{\chi}$. It is clear that once we know how the undetermined class of Morse meanders of type $\overline{\chi}$ is described, we are able to concatenate these with Morse meanders of type χ in order to describe combinatorially all Morse meanders. This is due to the fact that every Morse meander can be obtained by the right one-shift of meanders of the form $\mathcal{M}(A \mid B)$ with yet unknown requirements on A and B.

5 Connection Graphs of Type I, II, III and IV

The results of this section are motivated by Bernold Fiedler's work, [9], on global attractors of one-dimensional parabolic equations and the sixteen examples therein. Sturm global attractors with maximal Morse index two are completely characterized in [13] by inter alia introducing 1 -skeletons assigned to connection graphs. We will not consider the 1-skeletons here. We will introduce the definitions and results on connection graphs of type i, $i = I, II, III, IV$ determined from dissipative Morse meander permutations of Morse meanders of type i, which are by Theorem 4.2 pitchforkable and correspondingly the underlying connection graphs of type I, II, III, IV describe a subset of pitchforkable Sturm global attractors. Recall that a connection graph $\mathscr{C}$ is a directed graph in which the vertices stand for the equilibria and the arcs, i.e., directed edges, stand for heteroclinic connections between the equilibria. More precisely, in the PDE setting of Chapter 3 the connection graphs are constructed by the following theorems, cf. [9]:

Theorem. *Given any two equilibria v, w the permutation π associated to the nonlinearity f of PDE* (3.1) *determines, explicitly and constructively, whether or not v connects to w.*

Theorem (Cascading). *Let v, w denote any two equilibria such that for the Morse index holds: $\mathtt{i}(v) = \mathtt{i}(w) + n > \mathtt{i}(w)$. Then, $v \searrow w$ if and only if, there exists a sequence $e_0 = w, \ldots, e_n = v$ of equilibria such that $\mathtt{i}(e_k) = \mathtt{i}(w) + k$, and*

$$v = e_n \searrow e_{n-1} \searrow \ldots \searrow e_1 \searrow e_0 = w.$$

Before finishing the summary of the description of connection graphs of $\mathscr{A}$ we recall the formula for the zero-number of two equilibria v, w, cf. [9], Proposition 1.1.: For any pair v_m, v_n of equilibria, $m < n$, the zero number $z(v_n - v_m)$ is given by the following formula

$$z(v_n - v_m) = \mathtt{i}(v_m) + \frac{1}{2}\Big[(-1)^n \mathtt{sign}\Big(\pi^{-1}(n) - \pi^{-1}(m)\Big) - 1\Big] + \sum_{j=m+1}^{n-1} (-1)^j \mathtt{sign}\Big(\pi^{-1}(j) - \pi^{-1}(m)\Big), \tag{5.1}$$

where π is a dissipative Morse meander permutation and empty sums are zero, compare with Chapter 3. For the description of the connection graph of the attractor $\mathscr{A}$, it is important to know when a connection does not exist. The following theorem, cf. [3] and [4], gives such a characterization for non-existence of connections:

Theorem 5.1 (Blocking). *Let v, w be equilibria such that $\mathtt{i}(v) = \mathtt{i}(w) + 1$. Then v does not connect to w if at least one of the following two conditions is satisfied:*

a) *(Morse Blocking) $z(v - w) \neq \mathtt{i}(w)$, or*
b) *(Zero Number Blocking) there exists an equilibrium $\overline{w}$ such that $\overline{w}$ is between v and w, at $x = -1$, and $z(v - \overline{w}) = z(w - \overline{w})$.*

DOI 10.1515/9783110533026-006

To proceed further, the following theorem, cf. [9], completes the description of connection graphs of $\mathscr{A}$:

Theorem 5.2 (Liberalism). *Let v, w be equilibria with*

$$\mathtt{i}(v) = \mathtt{i}(w) + 1.$$

If the connection from v to w is not blocked, then v connects to w.

We recall from [29], that there is a slightly shorter way of describing the existence of heteroclinic connections between equilibria in the above setting by introducing the k-adjacency as follows:

Definition 5.1 (k-Ordered). *Let H be a subset of a phase space X, containing functions from $C^1[0,1]$ which satisfy Neumann boundary conditions. A pair $u_1, u_2 \in H$ with*

$$z(u_1 - u_2) = k$$

and all zeros of $u_1(x) - u_2(x)$ being simple is called k-ordered, *in symbols*

$$u_1 \prec_k u_2,$$

if we have $u_1(0) < u_2(0)$.

With the aid of the above definition in [29], M. Wolfrum has given an alternative way of formulating the above Theorems 5.1 and 5.2 by introducing the notion of k-adjacency.

Theorem 5.3 (k-Adjacency). *Two hyperbolic equilibria $v, w \in \mathscr{E}_f$ with $z(v - w) = k$ have a heteroclinic connection if and only if they are k-adjacent, in symbols*

$$v \prec\!\!\prec_k w,$$

i.e., if there is no third element $u \in H$ with

$$v \prec_k u \prec_k w.$$

Definition 5.2 (Initial and Final Vertices of a Connection Graph). *We call the two distinguished vertices of a connection graph $\mathscr{C}$, which are in correspondence with the initial and final vertices of the underlying Morse meander,* initial *and* final *and denote them by i and f respectively. For the connection graph $\mathscr{C}_A$ associated to a given Morse meander $\mathcal{M}$ we denote the initial and final points by i_A and f_A.*

Definition 5.3 (Symmetric Connection Graph). *We call a connection graph $\mathscr{C}$* symmetric *if there exists a graph involution τ, i.e., a graph automorphism of $\mathscr{C}$ such that*

$$\tau^2 = id,$$

which permutes the two distinguished vertices i and f.

5.1 Suspensions of Connection Graphs

The following material is based on the lecture notes on *infinite-dimensional dynamical systems* by Bernold Fiedler.

Construction 11 (Suspension of Connection Graphs). *Let $\mathscr{C}$ denote a connection graph with $2n + 1$ vertices labeled by $v_1, \dots, v_{2n+1}$. Suspend $\mathscr{C}$ as follows:*
1) *Insert two additional vertices i^* and f^* in the vertex set $V^{\mathscr{C}}$.*
2) *Connect any original sink $v \in \mathscr{C}$, i.e., equilibria with Morse index zero, with the new additional vertices i^* and f^* by a directed edge from v to i^* and f^*.*

Definition 5.4 (Relatively Symmetric Connection Graph). *We call a connection graph $\mathscr{C}$* relatively symmetric *if there exists a graph automorphism τ of $\mathscr{C}$ which permutes the two distinguished vertices i and f.*

By Construction 11, suspended connection graphs are symmetric, i.e., there exists a graph involution which permutes the two distinguished points. Each connection graph has exactly one initial and exactly one final vertex coming from the two end-vertices of each Morse meander. The product of two connection graphs $\mathscr{C}_1$ and $\mathscr{C}_2$ is defined as follows:

$$\begin{aligned} \mathscr{C}_1 \cdot \mathscr{C}_2 :\Leftrightarrow \text{ take the disjoint union of } \mathscr{C}_1 \text{ and } \mathscr{C}_2 \\ \text{and identify the final vertex of } \mathscr{C}_1 \text{ with the initial vertex of } \mathscr{C}_2. \end{aligned} \tag{5.2}$$

Two connection graphs $\mathscr{C}_1, \mathscr{C}_2$ are isomorphic as directed graphs, if there exists a graph isomorphism

$$\rho : \mathscr{C}_1 \to \mathscr{C}_2,$$

in symbols $\mathscr{C}_1 \cong \mathscr{C}_2$. Recall, that a graph isomorphism maps longest paths in $\mathscr{C}_1$ to longest paths in $\mathscr{C}_2$.

5.2 Construction of Connection Graphs of Type I, II, III and IV

Definition 5.5 (Connection Graph of Type Y). *A connection graph determined from Morse meanders of type*

$$Y, \quad Y \in \{I, II, III, IV, \dots\}$$

is called a connection graph of type Y. *We denote connection graph of type Y by $\mathscr{C}_Y$.*

In the subsequent sections we characterize combinatorially the four types of connection graphs determined by permutations of Morse meanders of type I, II, III and IV.

5.2.1 Connection Graphs of Type I

Definition 5.6 (Connection Graphs of Type I). *Let A be a bracket expression representing a block with K levels and let T_A be the canonical tree associated to A. We call connection graphs which correspond to right one-shifted elements*

$$\mathcal{M}(A \mid A[1])$$

connection graphs of type I.

Before proving Theorem 5.4, we need to prove the following proposition.

Proposition 5.1. *Let $a \in \mathbb{N}_{>0}$ be the number of arcs of a rainbow block and $\mathcal{M}(a \mid a[1])$ a right one-shifted element, which is a Morse meander of type I by Theorem 4.1. Then, the connection graph $\mathcal{C}_a$ associated to the Morse meander $\mathcal{M}(a \mid a[1])$ has the following properties:*

i) *(Graph Structure of $\mathcal{C}_a$) $\mathcal{C}_a$ consists of 2^a longest paths of length a. All paths start in v_a and terminate in i_a or f_a, i.e m_a is the only source and i_a, f_a are sinks.*
ii) *$\mathcal{C}_a$ is symmetric.*

Proof. By Proposition 4.1, the meander permutation of a right one-shifted element $\mathcal{M}(a \mid a[1])$ is given by

$$\pi_{CI}^{a} = [(2k \;\; 2(a-k+1))]_{k=1,\ldots,\lfloor \frac{a}{2} \rfloor} \text{ for } a > 1$$

and by $\pi = id$ for $a = 1$. By using Formula (5.1) on the zero-numbers of equilibria $v_j \in \mathcal{E} = \mathrm{V}^{\mathcal{M}(a|a[1])}$, $j = 1, \ldots, 2a$, we determine the zero-number matrix associated to $\mathcal{C}_a$:

$$M_a = \begin{pmatrix} 0 & 0 & 0 & \cdots & \cdots & \cdots & 0 & 0 & 0 \\ 0 & 1 & 1 & \cdots & \cdots & \cdots & 1 & 1 & 0 \\ 0 & 1 & 2 & \cdots & \cdots & \cdots & 2 & 1 & 0 \\ \vdots & \vdots & \vdots & \ddots & \vdots & \iddots & \vdots & \vdots & \vdots \\ \vdots & \vdots & \vdots & \cdots & a & \cdots & \vdots & \vdots & \vdots \\ \vdots & \vdots & \vdots & \iddots & \vdots & \ddots & \vdots & \vdots & \vdots \\ 0 & 1 & 2 & \cdots & \cdots & \cdots & 2 & 1 & 0 \\ 0 & 1 & 1 & \cdots & \cdots & \cdots & 1 & 1 & 0 \\ 0 & 0 & 0 & \cdots & \cdots & \cdots & 0 & 0 & 0 \end{pmatrix} \in Mat(2a+1, \mathbb{R}).$$

From the above zero-number matrix and by blocking conditions we have a source in m_a, whose degree of ingoing edges $\deg_{in}$ is zero and the degree of outgoing edges $\deg_{out}$ is two. Precisely, there is a directed edge from m_a to v_{a+1} and a directed edge from m_a to v_{a-1}. For the vertices v_{a+1} and v_{a-1} we have

$$\deg_{in}(v_j) = 1 \text{ and } \deg_{out}(v_j) = 2 \text{ for } j = a+1, a-1.$$

Moreover, the degrees of ingoing and outgoing edges of vertices

$$v_k \text{ and } v_{2a-k}, \; k = 2, \ldots, a-2,$$

equal to two. The distinguished points $i_a := v_1$ and $f_a := v_{2a+1}$ are sinks with $\deg_{in} = 2$. Therefore, the maximal distance from the source m_a to one of the distinguished nodes i_a or f_a equals to a which is the length of a longest path in $\mathcal{C}_a$. In particular, we have one possibility for going from the source m_a to v_{a+1} and from v_{a+1}

two possibilities each time we visit $a - 1$ vertices in order to arrive at the sink i_a. The same holds for visiting the nodes by starting at the source m_a in order to get to f_a. This shows that we have 2^a longest paths of length a. Property *ii)* follows by the fact that suspensions symmetrize connection graphs. □

Definition 5.7 (Opening Eye, Dilating Pupil). *A connection graph $\mathscr{C}_a$ determined from a permutation of a Chafee–Infante element*

$$\mathcal{M}(a \mid a[1]), \quad a > 1$$

is called an opening eye *of order a and a connection graph determined from the permutation of $\mathcal{M}(1 \mid 1[1])$ a* dilating pupil.

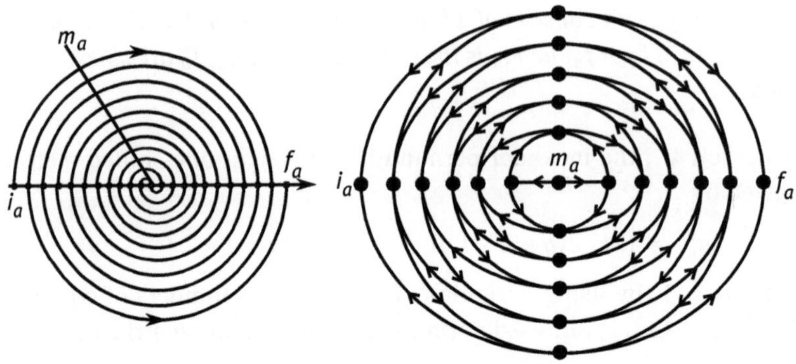

Fig. 5.1: Chafee–Infante element and the corresponding opening eye of order $a = 11$.

Note, that an opening eye

$$\mathcal{M}(a \mid a[1]), \quad a > 1$$

arises from an application of a-many suspensions of the dilating pupil $\mathcal{M}(1 \mid 1[1])$, in symbols $\mathcal{M}(a \mid a[1]) = \mathrm{sus}_a(\mathcal{M}(1 \mid 1[1]))$. Connection graphs of type I are suspended products of opening eyes. To be precise, connection graphs of type I are built as follows.

Construction 12 (Connection Graphs of Type I). *Let $\mathcal{M}(A \mid A[1])$ be given with a bracket expression A representing a block with K levels. The connection graph $\mathscr{C}_A$ is determined by following steps:*

Step 1. *We start with the leaves of the canonical tree T_A. Each leaf with weight $a_i > 1$ stands for an opening eye of order a_i and a leaf with weight 1 stands for a dilating pupil, see Definition 5.7.*

Step 2. *By gluing the distinguished points of opening eyes, respectively dilating pupils as defined in (5.2), we take the product of all opening eyes, respectively dilating pu-*

pils, if the children from which they were determined are joined by the same node in the next level.

Step 3. *We suspend the determined product of opening eyes corresponding to one root with weight r by following Construction 11 r-many times. The first pair of vertices* i_1^*, f_1^* *are inserted in such a way that* i_1^* *is above the product and* f_1^* *below. We draw a directed edge from the initial and final vertices of the product as well as from its glued vertices to* i_1^* *and* f_1^*. *We insert* i_2^* *on the left from the suspended product and* f_2^* *on the right and draw a directed edge from the vertices* i_1^*, f_2^* *which were inserted before. We proceed r − 2 times.*

Step 4. *We take products of determined connection graphs corresponding to one level in the canonical tree* T_A*, if there is one node in the level above joining the subtrees from which these connection graphs were determined.*

Step 5. *We are done by repeating the above steps until we reach the root of* T_A *indicating the final number of suspensions.*

Let

$$\mathcal{M}_1 := \mathcal{M}(1(1, 2(3(1, 3, 1), 2(3, 2, 3))) \mid 1(1, 2(3(1, 3, 1), 2(3, 2, 3))).$$

We illustrate Construction 12 by considering Morse meander $\texttt{shift}_{[1]}(\mathcal{M}_1)$ depicted in Figure 5.5, c). In Figure 5.2, we have written down the canonical tree associated to the

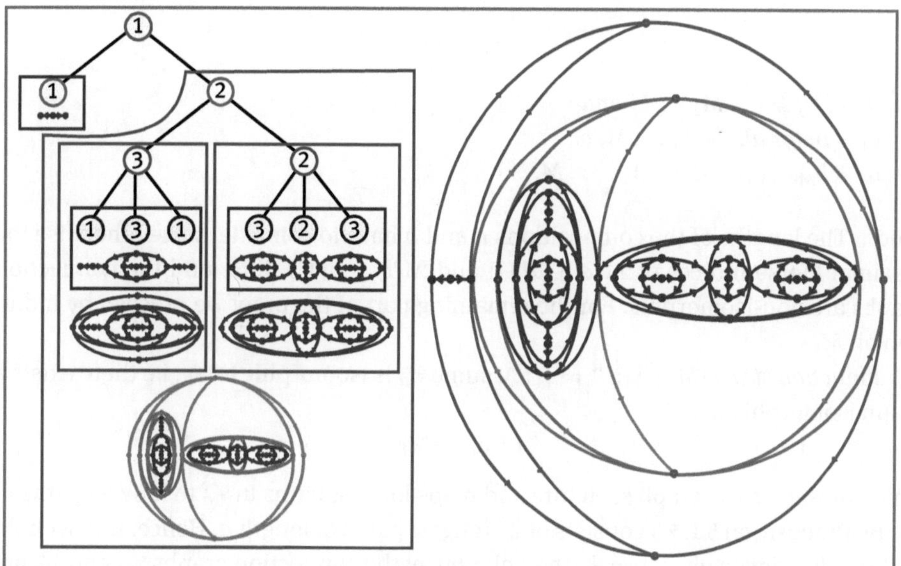

Fig. 5.2: Left: canonical tree and connection graphs built level-wise as described in Construction 12; right: complete connection graph of type I determined from the dissipative Morse meander permutation of $\texttt{shift}_{[1]}(\mathcal{M}_1)$; The dotted boxes around the levels and the corresponding connection graphs indicate the successive growing of the connection graph.

composition of the upper block represented by

$$1(1, 2(3(1, 3, 1), 2(3, 2, 3))),$$

as well as the corresponding connection graphs with respect to each level of the underlying canonical tree, see the dashed boxes, in order to demonstrate how the connection graph grows when we approach its root.

5.2.2 Isomorphism Class of Type I Connection Graphs

It is natural to formulate the following question: What are the necessary and sufficient condition for two products of opening eyes

$$\prod_{i=1}^{M} \mathcal{C}_{a_i} \text{ and } \prod_{i=1}^{M} \mathcal{C}_{b_i}$$

to be isomorphic? The following theorem answers this question.

Theorem 5.4. *Let $a = a_1, \ldots, a_{M_1}$ and $b = b_1, \ldots, b_{M_2}$ represent rainbow blocks. Let $\mathcal{M}(a \mid a[1])$ and $\mathcal{M}(b \mid b[1])$ be seaweed meanders, which are Morse by Theorem 4.1. Then, the associated connection graphs $\mathcal{C}_a$ and $\mathcal{C}_b$ are isomorphic, in symbols*

$$\mathcal{C}_a \cong \mathcal{C}_b$$

if and only if $M_1 = M_2 =: M$ and
1) *$a_i = b_i$ for all $i = 1, \ldots M$, or*
2) *$a_i = b_{M-i+1}$ for all $i = 1, \ldots, M$.*

Proof. The lengths of the compositions a and b have to coincide, since otherwise the maximal Morse indices of $\mathcal{M}(a \mid a[1])$ and $\mathcal{M}(b \mid b[1])$ differ and the connection graphs are non-isomorphic. For the remaining part of the proof we proceed by induction on M.

Induction start: $[M = 1]$; " $\Leftrightarrow$ " : Assume $\mathcal{C}_a$ is isomorphic to $\mathcal{C}_b$, i.e there exists a graph isomorphism

$$\iota : \mathcal{C}_a \to \mathcal{C}_b$$

which preserves the graph structure and maps longest paths in $\mathcal{C}_a$ to longest paths in $\mathcal{C}_b$. By Proposition 5.1, $\mathcal{C}_a$ consists of 2^a longest paths of length a. Hence, it must hold that $a = b$. Conversely, if $a = b$, then obviously the connection graphs $\mathcal{C}_a$ and $\mathcal{C}_b$ are isomorphic.

Induction assertion:

$$\prod_{i=1}^{M} \mathcal{C}_{a_i} \cong \prod_{i=1}^{M} \mathcal{C}_{b_i} \Leftrightarrow a_i = b_i \text{ or } a_i = b_{M-i+1}.$$

Induction hypothesis:

$$\prod_{i=1}^{M+1} \mathscr{C}_{a_i} \cong \prod_{i=1}^{M+1} \mathscr{C}_{b_i} \Leftrightarrow a_i = b_i \text{ or } a_i = b_{M-i+2}.$$

Induction step: $[M \rightsquigarrow M+1]$; " $\Leftrightarrow$ " : First note, that a graph isomorphism ι maps the distinguished points i_{a_1} and $f_{a_{M+1}}$ of $\mathscr{C}_{a_1} \cdot \ldots \cdot \mathscr{C}_{a_{M+1}}$ to distinguished points of $\mathscr{C}_{b_1} \cdot \ldots \cdot \mathscr{C}_{b_{M+1}}$. Namely, it holds

$$\iota(i_{a_1}) = \begin{cases} i_{b_1} & \text{if} \quad \iota(f_{a_{M+1}}) = f_{b_{M+1}} & (5.3a) \\ f_{b_{M+1}} & \text{if} \quad \iota(f_{a_{M+1}}) = i_1 & (5.3b) \end{cases}$$

and hence

$$\iota : \mathscr{C}_{a_1} \mapsto \mathscr{C}_{b_1} \quad \text{and} \quad \iota : \mathscr{C}_{a_{M+1}} \mapsto \mathscr{C}_{b_{M+1}} \quad \text{or}$$

$$\iota : \mathscr{C}_{a_1} \mapsto \mathscr{C}_{b_{M+1}} \quad \text{and} \quad \iota : \mathscr{C}_{a_{M+1}} \mapsto \mathscr{C}_{b_1}.$$

Since ι is a graph isomorphism, it preserves the length of the longest paths. By Proposition 5.1 and induction assertion we get the desired claim. □

Our aim for this section consists in the generalization of the above Theorem 5.4 for connection graphs corresponding to right one-shifted Morse meanders

$$\texttt{shift}(\mathcal{M}(A \mid A[1])),$$

where A represents a general block with K-many levels. We recall from Chapter 2 in the Notation 2, that for each bracket expression A representing a block with K-many levels we can associate the *canonical tree* T_A. The canonical tree

$$\mathrm{T}_A = (\mathrm{V}^{\mathrm{T}_A},\ \mathrm{E}^{\mathrm{T}_A},\ r,\ \omega : \mathrm{V}^{\mathrm{T}_A} \to \mathbb{N}_{>0})$$

is a rooted weighted tree with the root r and the weight function ω on the vertex set $\mathrm{V}^{\mathrm{T}_A}$. With respect to each level

$$k = 0, \ldots, K-1$$

of T_A, we define the linear order in the following way:

Definition 5.8 (Tree Order). *Let A and T_A be as above. A tree order $<$ is a linear order on each set of vertices*

$$\mathrm{V}_k^{\mathrm{T}_A} := \left\{ v \in \mathrm{V}^{\mathrm{T}_A} \mid \texttt{dist}(v, r) = k \right\},$$

of the kth-level of T_A such that if

$$x, y \in \mathrm{V}_k^{\mathrm{T}_A} \text{ and } x', y' \in \mathrm{V}_{k+1}^{\mathrm{T}_A} \text{ with } xx', yy' \in \mathrm{E}^{\mathrm{T}_A} \text{ then } x < y \text{ implies } x' < y'.$$

We say that $<$ defines a tree order on T_A.

Let $\mathsf{Ord}(\mathtt{T}_A)$ be the set of tree orders on $\mathtt{T}_A$.

Definition 5.9 (Equivalent Tree Orders). *Let* $<_1, <_2 \in \mathsf{Ord}(\mathtt{T}_A)$. *We call two tree orders* $<_1$ *and* $<_2$ *equivalent, in symbols*

$$<_1 \sim <_2,$$

if the connection graphs $\mathscr{C}_{(\mathtt{T}_A,<_1)}$ *and* $\mathscr{C}_{(\mathtt{T}_A,<_2)}$ *are isomorphic.*

In the sequel we denote the canonical trees by a triple $(\mathtt{T}_A, <, r)$ in order to specify the root r and tree order $<$ taken under consideration. Let $(\mathtt{S}_A, <, s)$ be a *maximal* subtree with root s of a tree $(\mathtt{T}_A, <, r)$ with root r, i.e., for all $x \in \mathtt{V}^{\mathtt{S}_A}$ holds

$$\mathtt{dist}(r, x) = \mathtt{dist}(r, s) + \mathtt{dist}(s, x).$$

We define a map $\mathtt{reflect}_{(\cdot)}$ on the set of tree orders $\mathsf{Ord}(\mathtt{T}_A)$ as follows:

$$\begin{aligned} \mathtt{reflect}_s : \mathsf{Ord}(\mathtt{T}_A) &\to \mathsf{Ord}(\mathtt{T}_A) \\ < &\mapsto <^{*,s}, \end{aligned}$$

where $<^{*,s}$ is the tree order obtained from $<$ by dualizing $<$ along the maximal subtree S_A with root s, i.e., for $x, y \in \mathtt{V}^{\mathtt{S}_A}$ holds

$$y <^* x \quad :\Leftrightarrow \quad x < y$$

Remarks 5.1.
i) *The group* Ref *generated by the maps* $\mathtt{reflect}_{(\cdot)}$ *is a Coxeter group. We denote the group operation of* Ref *by* '$\boldsymbol{.}$'.
ii) *When considering binary trees we can obtain all tree orders from a given one by a sequence of reflections (a sequence of applications of map* $\mathtt{reflect}_{(\cdot)}$ *) along its maximal subtrees.*

Example 5.1 (Reflection on a Maximal Subtree and 2 Isomorphic Connection Graphs). *Let*

$$A = a^0\Big(a_1^1(a_1^2, a_2^2),\ a_2^1(a_3^2(a_1^3, a_2^3, a_3^3), a_4^2, a_5^2(a_4^3, a_5^3))\Big)$$

represent a block with 4 levels and let $\mathtt{T}_A$ *be its associated canonical tree. We consider* $\mathtt{reflect}_{a_2^1}$ *on the maximal subtree* $(\mathtt{S}_A, <, a_2^1)$ *of the tree* $(\mathtt{T}_A, <, a^0)$:

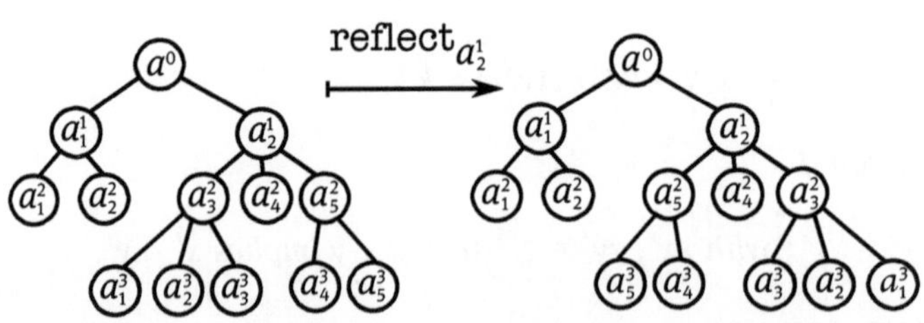

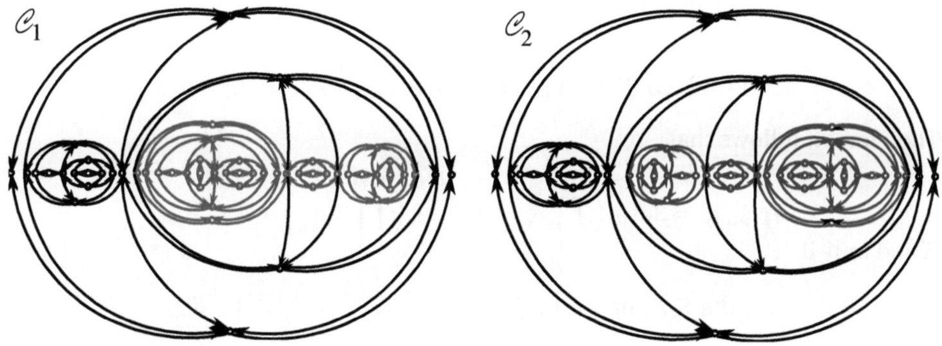

Fig. 5.3: An example of two isomorphic connection graphs.

The following two connection graphs in Figure 5.3 are isomorphic. In order to construct an isomorphism $\iota : \mathscr{C}_1 \to \mathscr{C}_2$, it suffices to relabel the vertices for the subgraph (blue) of $\mathscr{C}_2$ in the reversed order of the labeled vertices of the subgraph (red) of $\mathscr{C}_1$. All other labels of vertices remain the same. The canonical tree of the left connection graph $\mathscr{C}_1$ is given by to the maximal canonical subtree corresponding to the bracket expression $2(4(1,2,3),3,2(1,2))$ *we get the canonical tree of the right connection graph $\mathscr{C}_2$.*

The following theorem generalizes Theorem 5.4.

Theorem 5.5. *Let A be a bracket expression representing a block with K levels and T_A its canonical tree. Then, two connection graphs $\mathscr{C}_{(\mathrm{T}_A,<,r)}$ and $\mathscr{C}_{(\mathrm{T}_A,<',r)}$ are isomorphic if and only if*

$$<' = g. < \text{ for } g \in \mathrm{Ref}.$$

Proof. We proceed by strong induction on the number of levels K of a block. When there is only one level, i.e., $K = 1$, every canonical tree T_A consists only of the 0th-level, such that $g = id \in \mathrm{Ref}$ by Proposition 5.1. The case $K = 2$ is more compelling. Let

$$\hat{A} = a_0(a_1^1, \dots, a_{L(1)}^1) = a_0(A), \quad L(1) \in \mathbb{N}_{>1},$$

i.e., $\hat{A}$ represents a block with two levels, and $T_{\hat{A}}$ the corresponding canonical tree. Let $Set(\mathscr{C}, J)$ denote the set of connection graphs corresponding to a right one-shifted element $\mathcal{M}(A' \mid A'[1])$, where A' represent a block with J levels and let

$$\begin{aligned} \mathrm{Sus}_{a_0} &: Set(\mathscr{C}, K) \to Set(\mathscr{C}, K+1) \\ &\mathscr{C}_A \mapsto \mathscr{C}_{\hat{A}} = \mathscr{C}_{a_0(A)} =: \mathrm{Sus}_{a_0}(\mathscr{C}_A) \end{aligned}$$

be the suspension map of a connection graph $\mathscr{C}_A = \mathscr{C}_{T_A}$. By Theorem 5.4, it follows that

$$\mathscr{C}_{(T_A,<)} = \prod_{i=1}^{L(1)} \mathscr{C}_{a_i^1} \cong \prod_{i=1}^{L(1)} \mathscr{C}_{a_{\tau(i)}^1} = \mathscr{C}_{(T_A,<')}$$

if and only if

$$\tau = id \in S_{L(1)} \text{ or } \tau = \begin{pmatrix} 1 & 2 & 3 & \cdots & L(1)-1 & L(1) \\ L(1) & L(1)-1 & L(1)-2 & \cdots & 2 & 1 \end{pmatrix} \in S_{L(1)}.$$

Therefore, it follows that

$$\mathscr{C}(T_{\hat{A}},<,a_0) = \mathrm{Sus}_{a_0}\Big(\prod_{i=1}^{L(1)} \mathscr{C}_{a_i^1}\Big) \cong \mathrm{Sus}_{a_0}\Big(\prod_{i=1}^{L(1)} \mathscr{C}_{a_{\tau(i)}^1}\Big) = \mathscr{C}(T_{\hat{A}},<',a_0)$$

if and only if

$$\tau = id \in S_{L(1)} \text{ or } \tau = \begin{pmatrix} 1 & 2 & 3 & \cdots & L(1)-1 & L(1) \\ L(1) & L(1)-1 & L(1)-2 & \cdots & 2 & 1 \end{pmatrix} \in S_{L(1)}$$

which is equivalent to require

$$<' = g. < \text{ with } g = id \in \mathrm{Ref} \text{ or } g = \mathtt{reflect}_{a_0} \in \mathrm{Ref}.$$

i) *Induction assertion:* The theorem holds for a bracket expression A representing a block with at most K levels.
ii) *Induction hypothesis:* The theorem holds for a bracket expression $\hat{A}$ representing a block with $K+1$ levels.
iii) *Induction step:* $[K \curvearrowright K+1]$.

To iii): $[K \curvearrowright K+1]$. Let

$$\hat{A} = a_0(A_1^1, \ldots, A_{L(1)}^1), \quad L(1) \in \mathbb{N}_{>1}$$

be a bracket expression representing a block with $K+1$ levels and $\mathtt{T}_{\hat{A}}$ the canonical tree associated to $\hat{A}$. We first show " $\Rightarrow$ ". We consider the following for a connection graph $\mathscr{C}_{T_{\hat{A}}}$:

$$\mathscr{C}_{T_{\hat{A}}} = \mathrm{Sus}_{a_0}\Big(\prod_{i=1}^{L(1)} \mathscr{C}_{T_{A_i^1}}\Big).$$

For each $\mathscr{C}_{T_{A_i^1}}$, $i = 1, \ldots, L(1)$ holds: A_i^1 represent a block with at most K levels. Therefore, by induction assertion the theorem holds for each $\mathscr{C}_{T_{A_i^1}}$, $i = 1, \ldots, L(1)$. It remains to show that if

$$\mathscr{C}(T_{\hat{A}},<,a_0) = \mathrm{Sus}_{a_0}\Big(\prod_{i=1}^{L(1)} \mathscr{C}_{T_{A_i^1}}\Big) \cong \mathrm{Sus}_{a_0}\Big(\prod_{i=1}^{L(1)} \mathscr{C}_{T_{A_{\tau(i)}^1}}\Big) = \mathscr{C}(T_{\hat{A}},<',a_0)$$

it follows that $\tau = id \in S_{L(1)}$ or $\tau = \begin{pmatrix} 1 & 2 & 3 & \cdots & L(1)-1 & L(1) \\ L(1) & L(1)-1 & L(1)-2 & \cdots & 2 & 1 \end{pmatrix} \in S_{L(1)}$. Assume,

$$\mathrm{Sus}_{a_0}\Big(\prod_{i=1}^{L(1)} \mathscr{C}_{T_{A_i^1}}\Big) \cong \mathrm{Sus}_{a_0}\Big(\prod_{i=1}^{L(1)} \mathscr{C}_{T_{A_{\tau(i)}^1}}\Big),$$

which is equivalent to $\prod_{i=1}^{L(1)} \mathscr{C}_{T_{A_i^1}} \cong \prod_{i=1}^{L(1)} \mathscr{C}_{T_{A_{\tau(i)}^1}}$. By the property of a graph isomorphism ι the distinguished points of $\prod_{i=1}^{L(1)} \mathscr{C}_{T_{A_i^1}}$ are mapped as follows, compare with the proof of Theorem 5.4:

$$\iota(i_{A_1^1}) = \begin{cases} i_{A^1_{\tau(1)}} & \text{if} \quad \iota(f_{A^1_{L(1)}}) = f_{A^1_{\tau(L(1))}} \quad (5.4a) \\ f_{A^1_{\tau(L(1))}} & \text{if} \quad \iota(f_{A^1_{L(1)}}) = i_{A^1_{\tau(1)}} \quad (5.4b) \end{cases}$$

since the distinguished points are the only vertices with

$$\deg_{in} = 2 \text{ and } \deg_{out} = 0.$$

The length of a longest path in $\mathscr{C}_{T_{A_i^1}}$ equals to the sum over all maximal weights with respect to each level of $T_{A_i^1}$. For this it suffices to note, that in each level of $T_{A_i^1}$ except of the last one, the weights stand for the number of suspensions of a product of Chafee–Infante eyes. Let $\mathscr{C}_{T_{A_j^1}}$ be non-isomorphic. Otherwise we would end up with trivial permutations, namely of two isomorphic $\mathscr{C}_{T_{A_j^1}}$ and $\mathscr{C}_{T_{A_k^1}}$. Since ι is a graph isomorphism the lengths of longest paths, the number of paths and the sequence of degrees of ingoing and outgoing edges of each vertex of $\mathscr{C}_{T_{A_i^1}}$ must be preserved under ι. Since all $\mathscr{C}_{T_{A_i^1}}$ are symmetric, we obtain the desired claim. " $\Leftarrow$ " : Conversely, consider a canonical tree $T_{\hat{A}}$ with $K + 1$ levels:

$(T_{\hat{A}}, <, a_0) =$

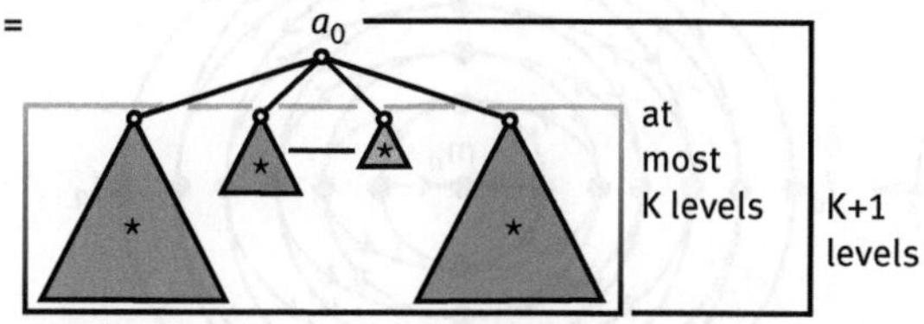

where the triangles stand for maximal subtrees $T_{A_1^1}, \ldots, T_{A^1_{L(1)}}$. By induction assertion the theorem holds for $T_{A_1^1}, \ldots, T_{A^1_{L(1)}}$.

Consider the same tree as before but with the reversed tree order $<'$, i.e.,

$(T_{\hat{A}}, <', a_0) =$

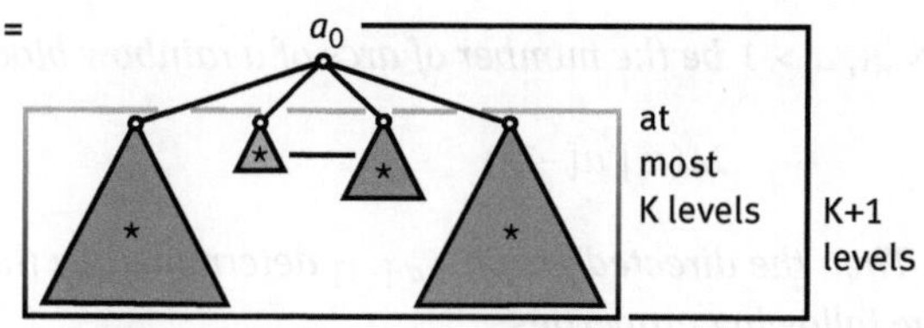

where the triangles stand for maximal subtrees $T_{A^1_{L(1)}}, \ldots, T_{A_1^1}$.

Since $a_0 > 0$, otherwise the number of levels of $T_{\hat{A}}$ would be less than $K + 1$, the connection graph $\mathscr{C}_{(T_{\hat{A}}, <, a_0)}$ is symmetric. Hence, both connection graphs $\mathscr{C}_{(T_{\hat{A}}, <, a_0)}$ and $\mathscr{C}_{(T_{\hat{A}}, <', a_0)}$ are isomorphic. □

5.2.3 Connection Graphs of Type II

In order to describe the connection graphs of type II, we consider directed graphs associated to left one-shifted elements $\mathcal{M}(A \mid A[-1])$, where A stands for a bracket expression representing a block with K levels. We are careful to point out that the directed graphs which may be obtained from the left one-shifted elements $\mathcal{M}(A \mid A[-1])$ only become connection graphs of Sturm global attractors after successive suspensions. This is due to the fact that the left one-shifted elements $\mathcal{M}(A \mid A[-1])$ are not Morse meanders. But may be recovered by applying suspensions as was described in the construction of Morse meanders of type II.

Definition 5.10 (Closing Eye, Shrinking Pupil). *Let $a \in \mathbb{N}_{>0}$, $a > 1$. We call a directed graph $\mathscr{C}_{a,[-1]}$ determined from permutations of left one-shifted elements $\mathcal{M}(a \mid a[-1])$* a closing eye *and $\mathscr{C}_{1,[-1]}$ determined from permutations of $\mathcal{M}(1 \mid 1[-1])$* a shrinking pupil.

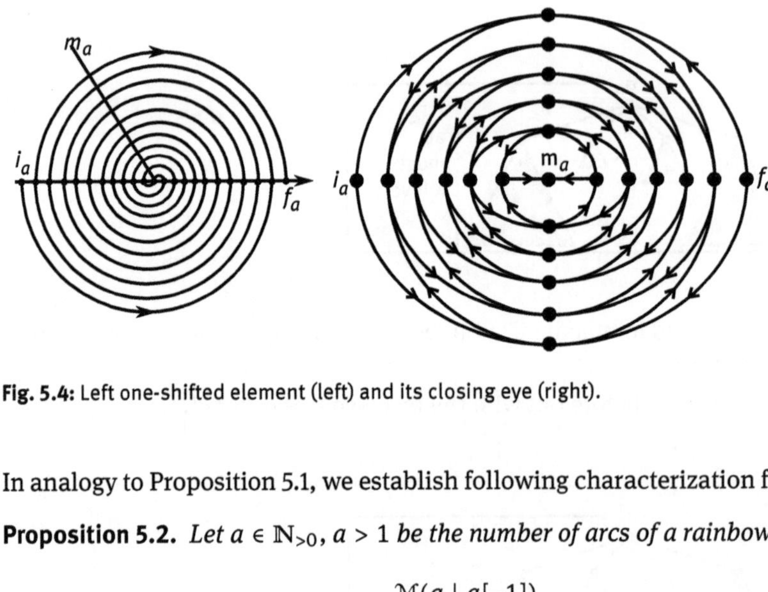

Fig. 5.4: Left one-shifted element (left) and its closing eye (right).

In analogy to Proposition 5.1, we establish following characterization for closing eyes.

Proposition 5.2. *Let $a \in \mathbb{N}_{>0}$, $a > 1$ be the number of arcs of a rainbow block and*

$$\mathcal{M}(a \mid a[-1])$$

a left one-shifted element. Then the directed graph $\mathscr{C}_{a,[-1]}$ determined by the permutation of $\mathcal{M}(a \mid a[-1])$ has the following properties:

i) *(Graph Structure of $\mathscr{C}_{a,[-1]}$) $\mathscr{C}_{a,[-1]}$ consists of 2^a longest paths of length a. All paths start in i_a or f_a and terminate in m_a.*
ii) *$\mathscr{C}_{a,[-1]}$ is symmetric.*

Proof. The left and right one-shifts defined for closed meanders are dual to each other, i.e., they are related by the identity

$$\mathtt{shift}_{[-1]}(\mathcal{M}(A \mid A)) = \mathtt{mirr}(\mathtt{shift}_{[1]}(\mathcal{M}(\hat{A}, \hat{A}))),$$

where $\mathtt{mirr}, \hat{\cdot}$ are as in Chapter 3 and A is a bracket expression representing a block. Therefore,

$$\mathcal{M}(a \mid a[-1]) = \mathtt{mirr}(\mathcal{M}(a \mid a[1])).$$

This translates into the reversal of all directed edges in $\mathscr{C}_a$ which is due to the multiplication of the corresponding zero-number matrix M_a in Proposition 5.1 by minus one. □

The next step in our consideration is the description of connection graphs determined by meander permutations corresponding to Morse meanders of type II. Recall from Chapter 4, that the class of Morse meanders of type II splits into A), B), C), D) and E). Correspondingly, the class of connection graphs of type II splits into A), B), C), D) and E). We therefore describe connection graphs of type II in A), B), C), D), E) separately in the construction given below.

Construction 13 (Connection Graphs of Type II). *Let*

$$A_i \neq 1,\ A_i \neq (1, \ldots, 1), \quad i = 1, \ldots, M$$

be a bracket expression representing a block with K levels and $\mathcal{M}(A_i \mid A_i[-1])$ a left one-shifted element. In order to construct connection graphs of type II we follow the steps listed below:

1) *We construct connection graphs $\mathscr{C}_{A_i}$ determined by permutations of right one-shifted elements $\mathcal{M}(A_i \mid A_i[1])$ by following Construction 12. We reorient obtained connection graphs by reversing the direction of all directed edges. This step is valid due to the identity obtained in (4.4). We denote directed graphs obtained in this way by $\mathscr{C}^*_{A_i}$.*
2) *We describe connection graphs corresponding to subclasses of the class of Morse meanders of type II:*
   A): *We take directed garphs $\mathscr{C}^*_{A_i}$ from 1) and insert a pair of vertices m_1 and m_2 such that m_1 is in one line left from the initial vertex of $\mathscr{C}^*_{A_i}$ and m_2 is in one line right from the final vertex of $\mathscr{C}^*_{A_i}$. We draw a directed edge from its initial vertex to m_1 as well as from its final vertex to m_2. This is in correspondence with the application of the map $\mathtt{mut}$ of the left one-shifted element $\mathcal{M}(A_i \mid A_i[-1])$. We denote the obtained graph by $\mathtt{Mut}(\mathscr{C}^*_{A_i})$. In order to perform suspensions of $\mathtt{Mut}(\mathscr{C}^*_{A_i})$ Γ_i-many times in correspondence with Γ_i-many suspensions of*

   $$\mathtt{mut}(\mathcal{M}(A_i \mid A_i[-1]))$$

   we follow Construction 11 and denote the obtained connection graphs by

   $$\mathtt{Sus}_{\Gamma_i}(\mathtt{Mut}(\mathscr{C}^*_{A_i})).$$

We multiply the obtained connection graphs given by

$$\mathtt{Sus}_{\Gamma_i}(\mathtt{Mut}(\mathscr{C}^*_{A_i}))$$

according to (5.2).

B): *We multiply* $\mathscr{C}^*_{A_i}$ *according to* (5.2). *Then, we construct*

$$\mathtt{Mut}\Big(\prod_{i=1}^{M}\mathscr{C}_{A_i}\Big)$$

same as for A) and suspend sufficiently many times.

C): *We take products of*

$$\mathtt{Mut}(\mathscr{C}^*_{A_i})$$

and connection graphs obtained in A) or B) and suspend according to the chosen number of suspensions of the underlying Morse meander, as described in Construction 8.

D): *We take products of connection graphs obtained in A), B), or C).*

E): *We suspend combinations of products obtained in D) arbitrary many times.*

We consider once again closed meander

$$\mathcal{M}_1 := \mathcal{M}(1(1,2(3(1,3,1),2(3,2,3)))\mid 1(1,2(3(1,3,1),2(3,2,3)))),$$

see Figure 5.5, b).

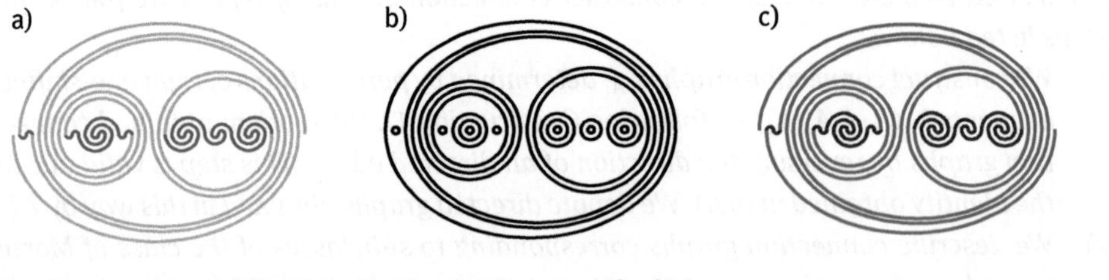

Fig. 5.5: a): Left one-shift of $\mathcal{M}_1$; b): $\mathcal{M}_1$; c) right one-shift of $\mathcal{M}_1$.

After applying the left one-shift on $\mathcal{M}_1$ we obtain a connected left one-shifted element $\mathtt{shift}_{[-1]}(\mathcal{M}_1)$ in Figure 5.5, a), which is not a Morse meander. We illustrate *1)* of Construction 13 by reversing all directed edges in the corresponding connection graph of type I in Figure 5.2.

In order to obtain the connection graph of type II from Figure 5.6, we need to apply the map Mut and then suspend the obtained graph at least 9 times. The connection graph obtained from a dissipative Morse meander permutation of

$$\mathtt{sus}_9(\mathtt{mut}(\mathtt{shift}_{[-1]}(\mathcal{M}_1)))$$

is depicted in Figure 5.7.

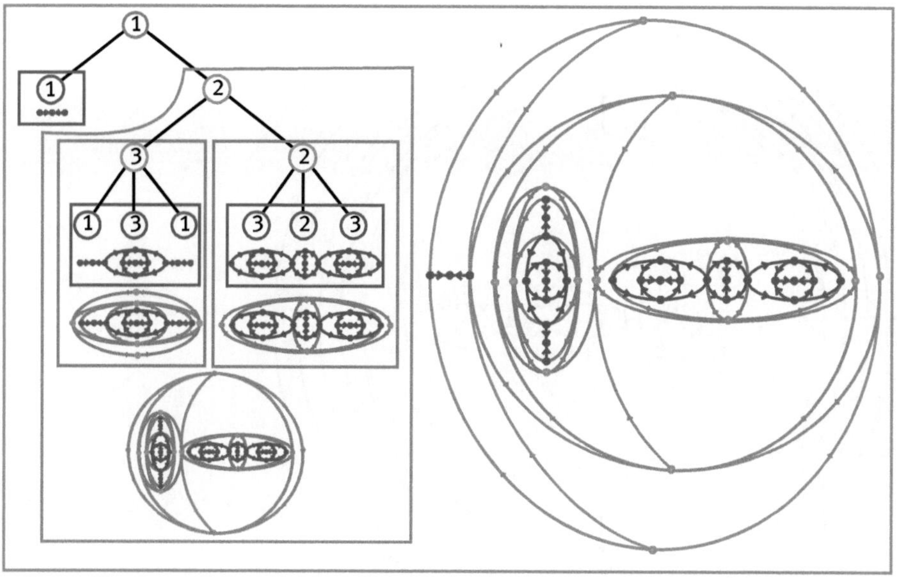

Fig. 5.6: Directed graph determined from the permutation of $\mathtt{shift}_{[-1]}(\mathcal{M}_1)$ by reversing all directed edges in Figure 5.2.

5.2.4 Isomorphism Class of Type II Connection Graphs

Let us consider isomorphism class of type II connection graphs in analogy to the isomorphism class of connection graphs of type I. For this reason we formulate and prove the following theorem:

Theorem 5.6. *Let $A \neq (1, \dots, 1), A \neq 1$ represent a block and*

$$\mathcal{M} = \mathtt{sus}_{\Gamma_A}(\mathtt{mut}(\mathcal{M}(A \mid A[-1])))$$

be Morse meander of type II. Let $\mathrm{T}^+ := \mathrm{T}^+_{\Gamma_A(1,A)}$ be the canonical tree of the upper block represented by $\Gamma_A(1, A)$ and $\mathrm{T}^- := \mathrm{T}^-_{\Gamma_A(A,1)}$ be the canonical tree of the lower block represented by $\Gamma_A(A, 1)$. Then two connection graphs of type II, $\mathscr{C}_{(\mathrm{T}^\pm,<_\pm,r)}$ and $\mathscr{C}_{(\mathrm{T}^\pm,<'_\pm,r)}$, are isomorphic if and only if

$$<'_+|_A = (g \centerdot <_+)|_A \quad \textit{for } g \in \mathrm{Ref},$$

i.e., the group of reflections restricted to the tree orders of the canonical tree $\mathrm{T}_A \subset \mathrm{T}^\pm$ operates on the tree orders assigned to the canonical tree $\mathrm{T}^\pm$ by fixing the order for the one single leaf contained in the first level and the subtree T_A.

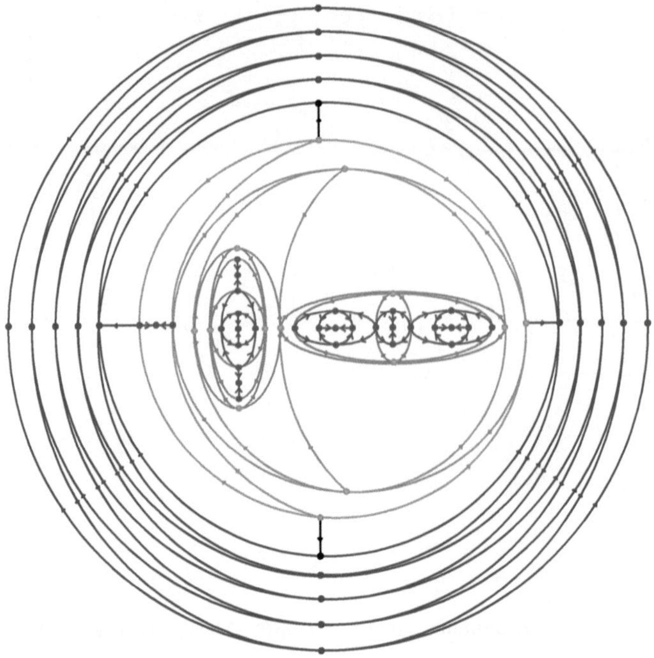

Fig. 5.7: Connection graph of type II determined from a dissipative Morse meander permutation of Morse meander $\mathtt{sus}_9(\mathtt{mut}(\mathtt{shift}_{[-1]}(\mathcal{M}_1)))$ being a Morse meander of type II; nine suspensions are depicted in green, connections determined from the mutation map $\mathtt{mut}$ in black and the connection graph corresponding to the above left one-shifted element in red and orange.

Proof. First note,

$$\begin{aligned}\mathcal{M} &= \mathtt{sus}_{\Gamma_A}(\mathtt{mut}(\mathcal{M}(A \mid A[-1])))\\ &= \mathtt{sus}_{\Gamma_A}(\mathtt{shift}_{[1]}(\mathcal{M}(1, A \mid A, 1))),\end{aligned}$$

which holds by (4.9). Since connection graphs of type II determined from the dissipative Morse meander permutation of $\mathcal{M}$ are obtained from Γ_A many suspensions of $\mathtt{Mut}(\mathscr{C}^*_{A,[-1]})$ which are constructed from connection graphs of type one $\mathscr{C}_{A,[1]}$ determined from

$$\mathtt{shift}_{[1]}(\mathcal{M}(A \mid A))$$

by reversing the edges and then applying $\mathtt{Mut}$, we may apply Theorem 5.5 to

$$\mathtt{Mut}(\mathscr{C}^*_{A,[-1]}).$$

This finishes the proof. □

5.2.5 Connection Graphs of Type III and IV

Construction 14 (Connection Graphs of Type III). *Recall, that Morse meanders of type III are suspensions of concatenated left and right one-shifted elements. For concatenation of a right one-shifted element with a left one-shifted element we introduced* $\circ_{\mathrm{R,L}}$ *and for concatenation of a left one-shifted element with a right one-shifted element:* $\circ_{\mathrm{L,R}}$. *After concatenations we end up with elements which are composed by*

$$\mathtt{mut}(M(A_i \mid A_i[-1])), A_i \neq 1, \; A_i \neq (1, \dots, 1)$$

and right one-shifted elements $M(A_j \mid A_j[1])$. *To obtain the corresponding connection graphs we take products of*

$$\mathtt{Mut}(\mathscr{C}^*_{A_i}) \text{ and } \mathscr{C}_{A_i},$$

as was described in the previous sections, in Constructions 12 and 13. We then suspend sufficiently many times.

Construction 15 (Connection Graphs of Type IV). *We define connection graphs of type IV as products of at least two different types of connection graphs of type I, II or III. We may suspend the resulting connection graph arbitrarily often.*

Remark. *Since connection graphs of type III and IV are built by using multiplication of right and left one-shifted elements our results in Theorem 5.5 and 5.6 hold for the factors.*

5.3 Connection Graphs of Different Types Are Non-Isomorphic

We prove the following theorem by using results of the previous sections. Let Con_l denote the class of connection graphs of type l, $l \in \{I, II, III, IV\}$.

Theorem 5.7. *Let* $\mathscr{C}_l \in Con_l$ *and* $\mathscr{C}_k \in Con_k$, $l, k \in \{I, II, III, IV\}$ *such that* $k \neq l$, *i.e.,* $\mathscr{C}_l$ *and* $\mathscr{C}_k$ *are two connection graphs determined by dissipative Morse meander permutations of Morse meanders of different types I, II, III, IV. Then, connection graphs of different types* $\mathscr{C}_l$ *and* $\mathscr{C}_k$ *for* $l \neq k$ *are non-isomorphic as directed graphs.*

Proof. Since the sets of Morse meanders of type I, II, III and IV are disjoint, which holds by Theorem 4.2, we have no connection graphs of type X which was determined by a Morse meander of type Y, $X \neq Y$, $X, Y \in \{I, II, III, IV\}$. Hence, there is no connection graph $\mathscr{C}_X$ which is of two types.

Let $\mathscr{C}_I \in Con_I$ be a connection graph of type I and $\mathscr{C}_{II} \in Con_{II}$ be a connection graph of type II. Since $\mathscr{C}_I$ and $\mathscr{C}_{II}$ contain non-isomorphic subgraphs, namely $\mathscr{C}_I$ contain opening and $\mathscr{C}_{II}$ closing eyes, it follows that $\mathscr{C}_I$ and $\mathscr{C}_{II}$ are non-isomorphic as directed graphs. For opening and closing eyes, see Definitions 5.7 and 5.10. Opening and closing eyes are non-isomorphic, since we obtain a closing eye by converting the source m of an opening eye into a sink and the sinks i, f of an opening eye into sources, which is by Propositions 5.1 and 5.2. Therefore, connection graphs of type I are non-isomorphic with connection graphs of type II.

Since connection graphs of type III are suspensions of products of opening and closing eyes and connection graphs of type IV also contain both opening and closing eyes, both types III and IV are non-isomorphic with connection graphs of type I and II. It remains to show that connection graphs of type III and IV are non-isomorphic. Connection graphs of type III are suspensions of concatenated opening and closing eyes, whereas connection graphs of type IV are products of at least two different types of connection graphs. Since connection graphs of type III do not contain products of connection graphs of type I and II or are multiplied with connection graphs of type I or type II, which holds for connection graphs of type IV, a connection graph $\mathscr{C}_{III}$ of type III and a connection graph $\mathscr{C}_{IV}$ of type IV are non-isomorphic. □

We depict some examples of connection graphs of type I, II, III and IV.

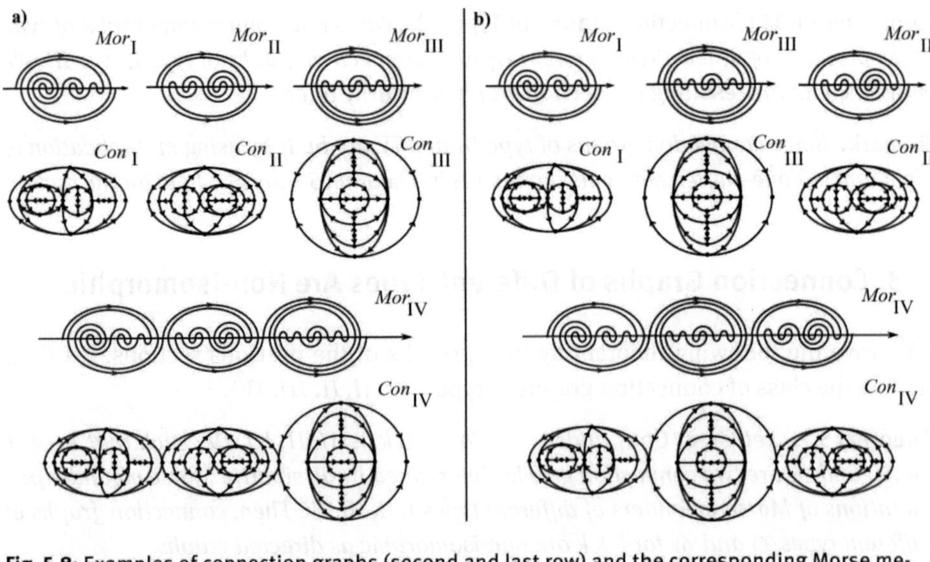

Fig. 5.8: Examples of connection graphs (second and last row) and the corresponding Morse meanders (first and third row) of four different types.

Our results of the previous sections are illustrated in Figures 5.9 and 5.10. We would like to put a special attention to the four connection graphs in line 27, 28, 29 and 30 in Figure 5.10. They are isomorphic by Theorem 5.5 and the fact that multiplication with dilating pupil from the left, respectively from the right, does not violate isomorphy. The number of non-isomorphic connection graphs with eleven vertices is 28. This is half of the determined total number of non-isomorphic connection graphs, cf. [9] in Table 2.1.

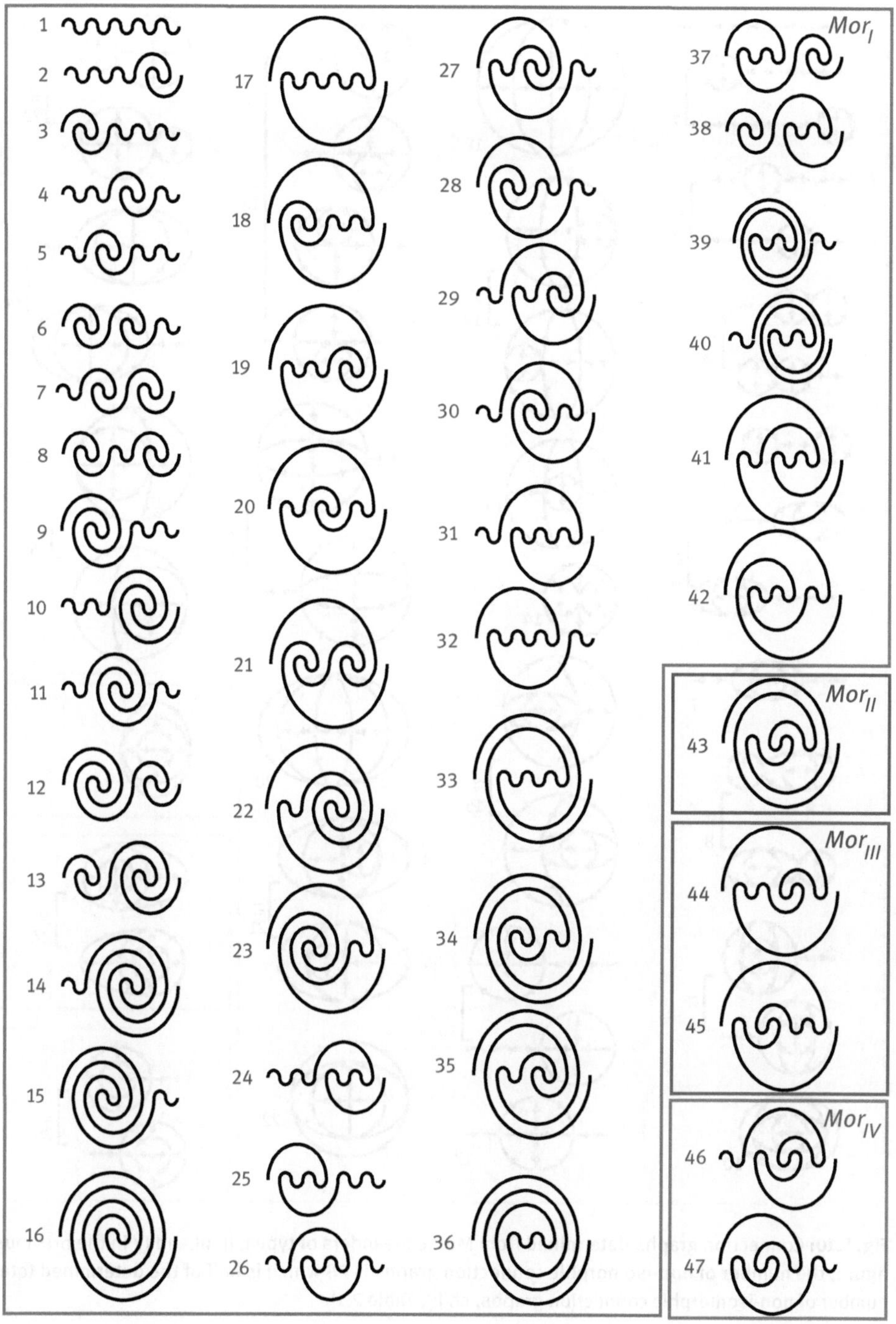

Fig. 5.9: Morse meanders of type I, II, III and IV as introduced in Chapter 4 with eleven transverse intersections. The horizontal oriented line has been omitted.

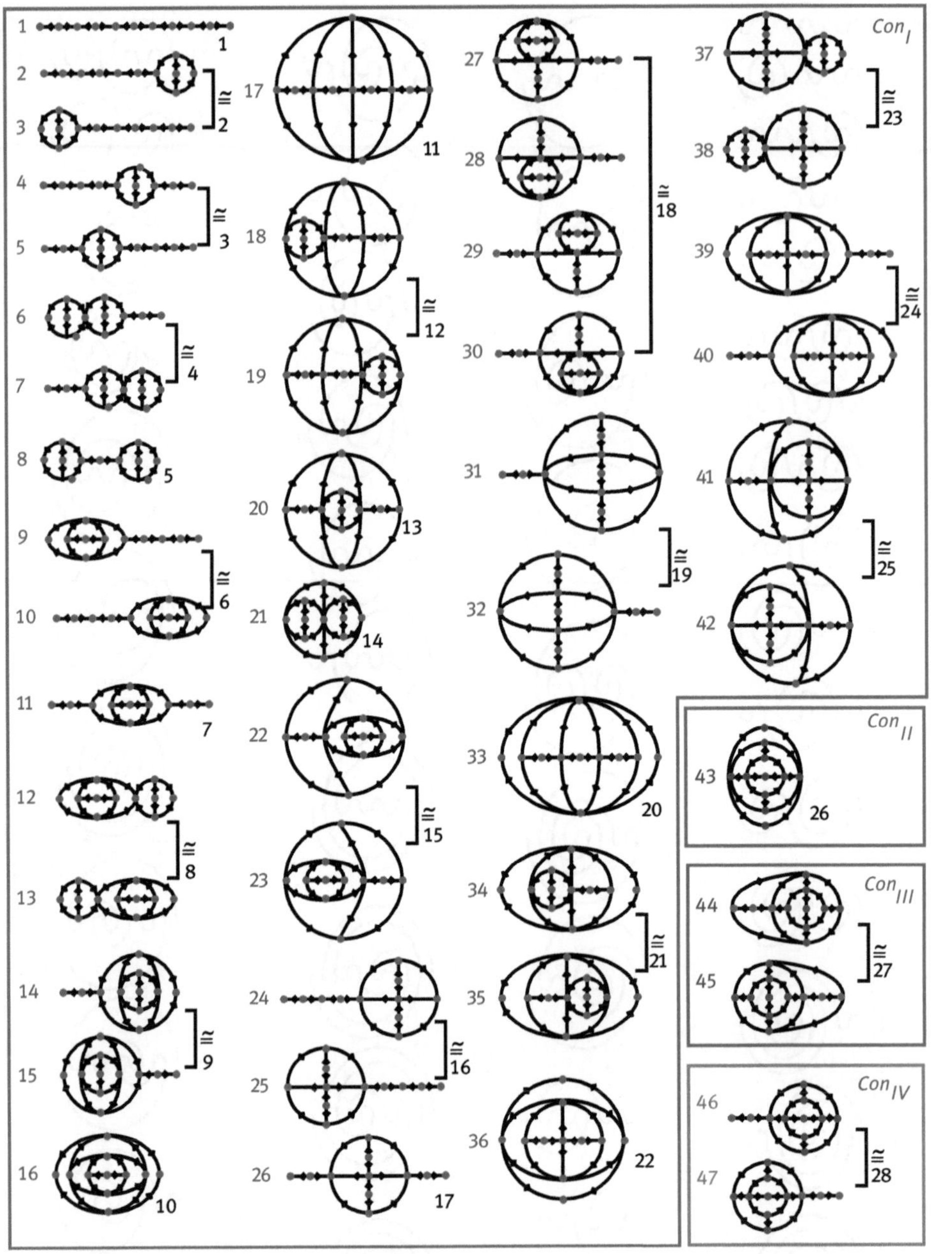

Fig. 5.10: Connection graphs determined from Morse meanders of type I, II, III and IV in the previous figure; the number of non-isomorphic connection graphs is 28 which is half of the determined total number of non-isomorphic connection graphs, cf. [9, Table 2.1].

6 Meanders and the Temperley–Lieb Algebra

6.1 From Open Meanders to Closed Meanders

We shall briefly introduce two ways of how to obtain *closed* meanders from *open* ones, since the closure of words expressed in terms of the generators of the Temperley–Lieb algebra represent closed meanders. A meander is considered to be *open* if there exist vertices of degree one, i.e., with only one neighbor. If all vertices have degree two, we call such meanders *closed*. It is clear that with this definitions collapsed meanders are i) open (if all connected components are paths), ii) open and closed (if all components are paths and cycles), or iii) closed (if all connected components are cycles). From Chapter 2, we know that meanders are closed before the collapse. Therefore, the concept of the collapse naturally yields the switching between open and closed meanders. In [10], the authors present the possibility of how to close a Morse meander, which is different from the closing procedure obtained by the collapse. To be precise, the closing in [10] of a Morse meander can be realized by the following map:

$$\texttt{cl} : \mathscr{M}or^{coll} \to \mathscr{M}$$
$$\mathcal{M}^{coll}(2A_1, \dots, 2A_N, 1 \mid 1, 2B_1, \dots, 2B_M)) \mapsto \mathcal{M}(A_1, \dots, A_N, 1 \mid 1(B_1, \dots, B_M)),$$

or equivalently,

$$\texttt{cl} : \mathscr{M}^{SW \rightsquigarrow NE} \to \mathscr{M}$$
$$\texttt{shift}_{[1]}(\mathcal{M}(A_1, \dots, A_N \mid B_1, \dots, B_M)) \mapsto \mathcal{M}(A_1, \dots, A_N, 1 \mid 1(B_1, \dots, B_M)),$$

where in the preimage A_l, $l = 1, \dots, N$ stands for positive number of nodes in the lth-upper block[1] and B_k, $k = 1, \dots, M$ for positive number of nodes in the kth-lower block. In the image of `cl` the numbers stand for the number of arcs.

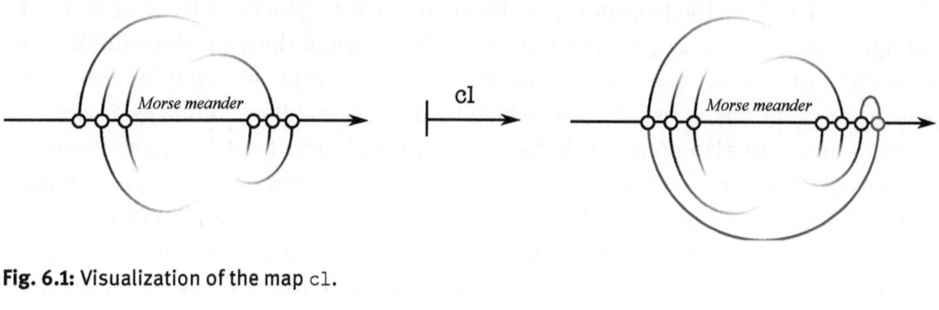

Fig. 6.1: Visualization of the map `cl`.

1 Recall that if the upper block A_l is non-rainbow, then A_l is a bracket expression in the Notation 1. The same holds for the lower blocks.

DOI 10.1515/9783110533026-007

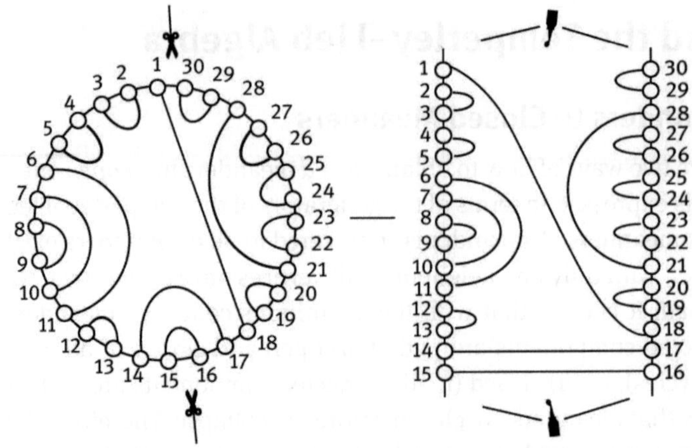

Fig. 6.2: An example of a strand diagram of order 15; left: in an open disc, right: in a rectangle. The scissors indicate where to cut the open disc.

6.2 From Blocks to Strand Diagrams

Following [21], we define the so called *strand diagrams*, which are used to pictorially represent the elements of the Temperly–Lieb algebra defined later on in the next section.

Definition 6.1 (Strand Diagram, [21]). *A strand diagram of order n, is a configuration of n pairwise non-intersecting curves, which are called strands, in an open disc, that connect $2n$ endpoints on the boundary of the disc. These points are labelled cyclically by $\{1, \dots, 2n\}$.*

Strand diagrams of order n may be represented in rectangles. The intersection points on the left side of the rectangle are labelled by $1, \dots, n$ and on the right side by $2n, \dots, n + 1$, from top to bottom, see Figure 6.2. Given a block $\mathscr{B}$ in $\mathbb{E}^+$ or $\mathbb{E}^-$ on the straight line $l = \mathbb{E}^+ \cap \mathbb{E}^-$, we can represent it in a strand diagram since the number of vertices of $\mathscr{B}$ is even. Note, that for the collapsed blocks, pictorial representation via strand diagrams as defined above is not valid, since collapsed blocks do not only contain arcs, but also points or forks, i.e., branched arcs. Let $1, \dots, 2n$ denote the vertices of $\mathscr{B}$. By cutting the line l between the vertex n and $n + 1$ and placing them parallel next to each other, we obtain the strand diagram in a rectangle corresponding to the block $\mathscr{B}$. Note, that since we do not rotate the two straight lines after the cutting, the labels of the vertices run from 1 to n on the left and from $2n$ to $n + 1$ on the right, from top to bottom. Conversely, we can glue the two parallel lines of a strand diagram in order to obtain a block on one straight line. Figure 6.4 on the right serves for an illustration. In comparison, if we are given strand diagrams in open discs then the order of cutting and gluing is reversed as indicated in Figure 6.4 on the left. Since we only cut

Fig. 6.3: Concatenation of two strand diagrams in rectangles (left column), respectively open discs (right column).

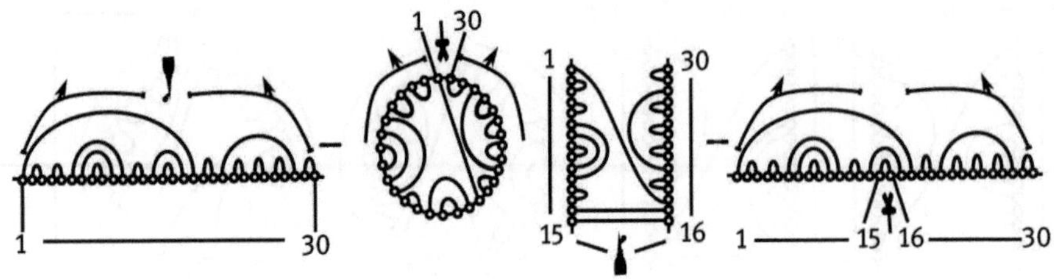

Fig. 6.4: Relating blocks to strand diagrams and vice versa, illustrated on a selected example.

or glue the straight lines and open discs, these deformations are independent from a given block. Strand diagrams and blocks are therefore in one-to-one correspondence, cf. [21].Given two strand diagrams we can concatenate (consult Figure 6.3) them by following:

Construction 16 (Concatenation of Two Strand Diagrams). *In order to multiply two elements* $\mathrm{e}, \mathrm{e}' \in \mathfrak{TL}_n(q)$, *we concatenate the corresponding strand diagrams of* e *and* e' *as follows:*

1) *We identify the right line of the strand diagram* e *with the left line of the strand diagram* e'.
2) *We identify the endpoint* i, $1 \leq i \leq n$ *in* e' *with the endpoint* $n - i + 1$ *in* e.
3) *Each time a closed loop is produced, we delete it and write a multiplicative factor* q *in front of the product* ee'.

6.3 Temperley–Lieb Algebra

From now on, we recall the definition of the braid group.

Definition 6.2. *The braid group* $\mathcal{B}_n$ *is defined by the generators*

$$\mathrm{g}_1, \ldots, \mathrm{g}_{n-1}$$

obeying the following relations:

$$\mathrm{g}_i\mathrm{g}_j = \mathrm{g}_j\mathrm{g}_i, \quad |i-j| > 1 \tag{6.1}$$

$$\mathrm{g}_i\mathrm{g}_{i+1}\mathrm{g}_i = \mathrm{g}_{i+1}\mathrm{g}_i\mathrm{g}_{i+1}, \quad i \in \{1, \ldots, n-2\}. \tag{6.2}$$

Definition 6.3. *The* Temperley–Lieb Algebra *of order n in the indeterminate q, denoted by* $\mathfrak{TL}_n(q)$, *is defined by the generators* $1, \mathrm{e}_i, i \in \{1, \ldots, n-1\}$ *fulfilling the following relations:*

$$\mathrm{e}_i^2 = q\mathrm{e}_i, \quad i = 1, \ldots, n \tag{6.3}$$

$$\mathrm{e}_i\mathrm{e}_j = \mathrm{e}_j\mathrm{e}_i, \quad \text{if } |i-j| > 1 \tag{6.4}$$

$$\mathrm{e}_i\mathrm{e}_{i\pm1}\mathrm{e}_i = \mathrm{e}_i, \quad i = 1, \ldots, n-2 \text{ if “}+\text{” and } i = 2, \ldots, n-1 \text{ if “}-\text{”}. \tag{6.5}$$

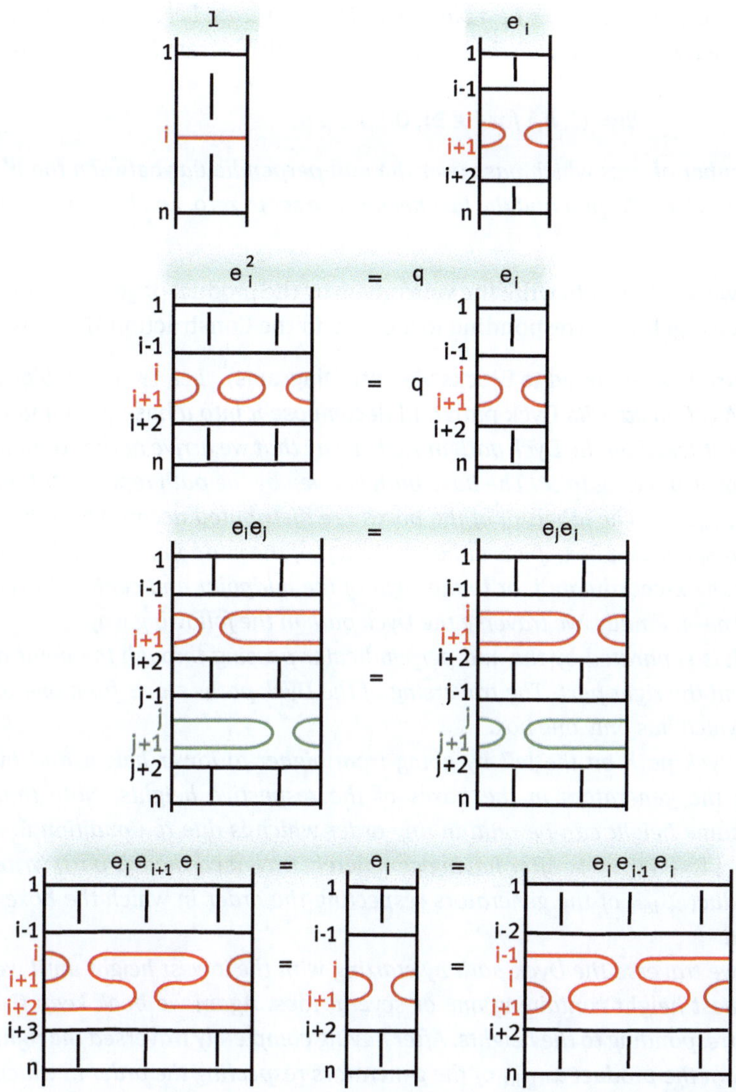

Fig. 6.5: Generators of the Temperley–Lieb algebra of order n with respective relations.

The generators

$$1, e_1, \ldots, e_n \text{ of } \mathfrak{TL}_n(q)$$

are represented by the strand diagrams depicted in Figure 6.5. We briefly recall from [21] how we obtain a word, i.e, a product of generators of Temperley–Lieb algebra $\mathfrak{TL}_n(q)$ assigned to blocks by using the so called *path diagrams* which are also called *Dyck* paths.

Definition 6.4 (Path Diagram). *Let $\mathscr{B}$ be a block with $2n$-many arcs. Let the corresponding vertices be ordered from v_1 to v_{2n} on the straight line. A* path diagram *is a graph in* $\mathbb{R}^2$ *on the vertices*

$$\vartheta_i = (i, h_i) \text{ for } i \in \mathbb{N}, 0 \le i \le 2n,$$

where h_i is the number of arcs which pass over the mid-perpendicular between the ith and $(i+1)^{st}$ vertices of $\mathscr{B}$. The first and the last heights are set to zero, i.e., $h_0 = h_{2n} = 0$.

By following [21], we recall how to write the word given by the product of generators of the Temperley–Lieb algebra, corresponding to a block, in the Construction 17 below.

Construction 17 (Words Associated to Blocks via Path Diagrams). *Let $\mathscr{B}$ be a block given by $A_1, \dots, A_N$. Consider its* Dyck path *and decompose it into a* base path *and a collection of* boxes *stacked on the Dyck path in such a way that we arrive at the contour of the Dyck path corresponding to $\mathscr{B}$. The base path is given by the path representation of $\mathfrak{I}_n^k = \mathrm{e}_1\mathrm{e}_3\mathrm{e}_5 \dots \mathrm{e}_{n-k-1}$. The labels e_i of the boxes are distributed according to their positions: We write e_i inside a box, if it is either to the left of the midpoint – given by n – and centered over the x-coordinate i, or to the right of the midpoint and centered over the x-coordinate $2n - i$. Finally, we traverse the Dyck path in the following way:*

1) *The Dyck path is separated by the mid-perpendicular passing through the point n into the left and the right part. The traversing of the Dyck paths starts from one of the left tops, which has only one box.*
2) *Traverse the Dyck path on the left by going from higher to lower height and by book keeping the generators in the boxes of the respective heights. Note that boxes at the same height can be read in any order which is due to Condition 6.5: $\mathrm{e}_i\,\mathrm{e}_j = \mathrm{e}_j\mathrm{e}_i$, $|i - j| > 1$. After having completely traversed the left part, write down the product $\mathfrak{P}_{left}$ of the generators respecting the order in which the boxes were traversed.*
3) *On the right, we traverse the Dyck path by starting with the lowest height until we reach the highest height containing one or several tops. Again we book keep the generators corresponding to the heights. After having completely traversed the right part, write down the product $\mathfrak{P}_{right}$ of the generators respecting the order in which the boxes were traversed.*
4) *Between the left and the right part of the Dyck path we write the product of generators corresponding to $\mathfrak{I}_n^k$, i.e.,*

$$\mathrm{e}_1\mathrm{e}_3\mathrm{e}_5 \dots \mathrm{e}_{n-k-1},$$

where k signifies the number of intersections of the mid-perpendicular through n and the arcs. The word $\mathcal{W}$ corresponding to the block $\mathscr{B}$ is given by multiplying $\mathfrak{P}_{left}$, $\mathfrak{I}_n^k$ and $\mathfrak{P}_{right}$, i.e.,

$$\mathcal{W} = \mathfrak{P}_{left}\mathfrak{I}_n^k\mathfrak{P}_{right}.$$

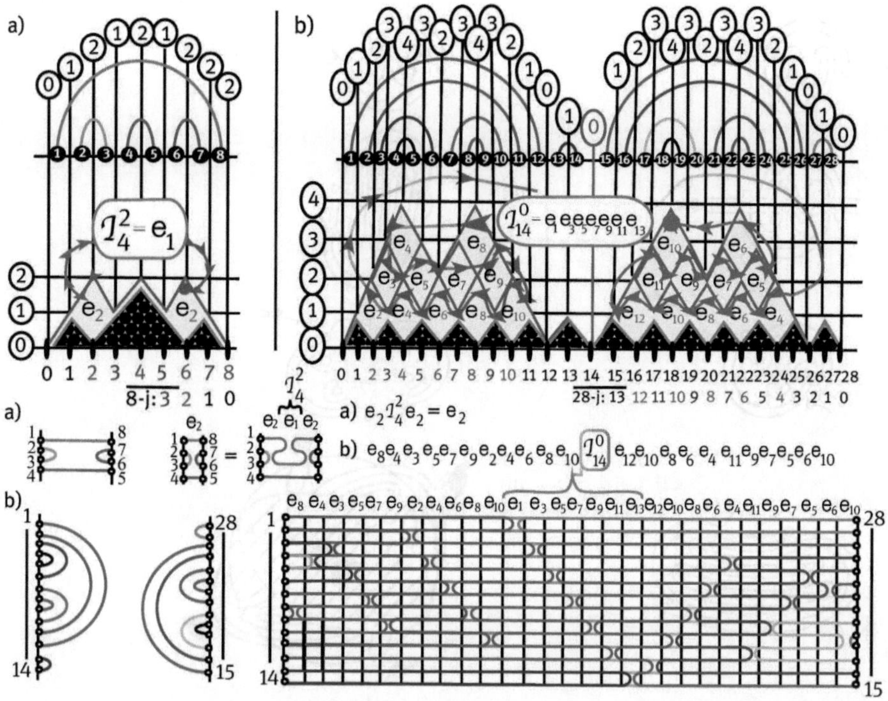

Fig. 6.6: Illustration of Construction 17 demonstrated on two examples in a) and b); base paths are indicated by light blue contours inside the Dyck paths; the determined words are confirmed by the strand diagrams depicted below.

In Figure 6.6 on the left, the block is given by $1(1,1,1)$. Note, that its corresponding word $\mathtt{e}_2\mathtt{e}_1\mathtt{e}_2$ may be simplified to e_2 by condition 6.5. In Figure 6.6 on the right, we have depicted the blocks given by

$$2(4,4),\ 2,\ 2(4,4),\ 2$$

and its corresponding path diagram. The associated word reads as follows:

$$\begin{aligned} \mathcal{W} &= \mathfrak{P}_{left}\mathfrak{I}^0_{14}\mathfrak{P}_{right} \\ &= (\mathtt{e}_8\mathtt{e}_4\mathtt{e}_3\mathtt{e}_5\mathtt{e}_7\mathtt{e}_9)(\mathtt{e}_1\mathtt{e}_3\mathtt{e}_5\mathtt{e}_7\mathtt{e}_9\mathtt{e}_{11}\mathtt{e}_{13})(\mathtt{e}_4\mathtt{e}_6\mathtt{e}_8\mathtt{e}_{10}\mathtt{e}_{12}\mathtt{e}_{11}\mathtt{e}_9\mathtt{e}_7\mathtt{e}_5\mathtt{e}_6\mathtt{e}_{10}). \end{aligned}$$

By closing the strand diagrams, that is by considering them either on cylinders (when we consider the strand diagrams in rectangles) or on spheres (when we work with strand diagrams in open discs) and by connecting the ith vertex with the $n-i+1^{st}$ vertex (compare with the definition of *rainbow* blocks), we obtain closed meanders as indicated in Figure 6.7 on cylinders or spheres. The number of connected components is then determined via the application of the following family of maps

$$\mathtt{trace}_n : \mathfrak{TL}_n(q) \to \mathbb{C}(q),\ for\ n \geq 1.$$

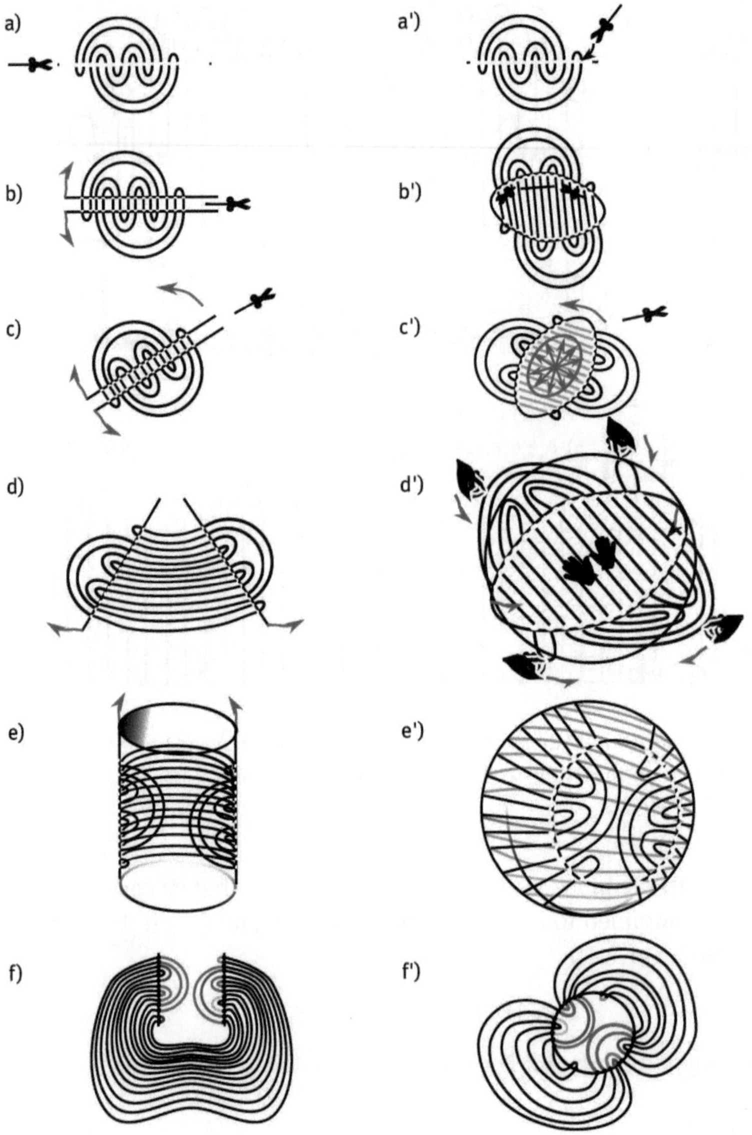

Fig. 6.7: Closure; a): cutting through the strip with a closed Morse meander; b): finishing the cutting and connecting the vertices doubled after the cutting; c): counter-clockwise rotation of b) by 90°; d): rotation of two parallel strips by 180°; e): stretching of the rotated d) onto the cylinder; f): planar drawing of e); a'): cutting through the middle of the strip between the first and the last vertex; b'): connecting the doubled vertices in the interior of the circle as indicated; c'): enlarging the interior part; d'): pushing the interior middle part down and pulling the strands on top of the sphere as indicated in e'); f'): planar drawing of e').

Definition 6.5 (Markov Trace, Trace). *A family of functions* $\mathtt{trace}_n$ *is called a* Markov trace *if it fulfills the following three properties:*

i) $\mathtt{trace}_n$ *is linear,*
ii) $\mathtt{trace}_n(\mathtt{e}\mathtt{e}') = \mathtt{trace}_n(\mathtt{e}'\mathtt{e})$ *for all* $\mathtt{e}, \mathtt{e}' \in \mathfrak{TL}_n(q)$*,*
iii) $\mathtt{trace}_{n+1}(\mathtt{e}\mathtt{e}_n) = \mathtt{trace}_n(\mathtt{e})$ *if* $\mathtt{e} \in \langle 1, \mathtt{e}_1, \ldots, \mathtt{e}_{n-1}\rangle$*.*

We call $\mathtt{trace}_n$ *satisfying only i) and ii) a* trace.

Given a word $\mathcal{W}$ associated to the lower and upper blocks of a meander we can determine the number of connected components by evaluating $\mathtt{trace}_n$ in $\mathcal{W}$. Precisely, the number of connected components is counted by the exponent of the multiplicative factor $q \in \mathbb{C}(q)$, which is by the following proposition, cf. [21].

Proposition 6.1. *Let* $\mathrm{L}(\mathcal{W})$ *denote the number of loops in the closure of* $\mathcal{W}$*. The family of functions* $\mathtt{trace}_n$ *defined on strand diagrams by* $\mathtt{trace}_n(\mathcal{W}) = q^{\mathrm{L}(\mathcal{W})}$ *and extended linearly to* $\mathfrak{TL}_n(q)$ *is the Markov trace such that* $\mathtt{trace}_1(1) = q^1$*.*

Note, that $\mathtt{trace}_n(1) = q^n$ which follows inductively. The induction start holds by definition: $\mathtt{trace}_1(1) = q^1$. The induction step ($n - 1 \rightsquigarrow n$) holds by the following calculation:

$$\begin{aligned}
\mathtt{trace}_n(1) &\overset{\text{def. 6.5,iii)}}{=} \mathtt{trace}_{n+1}(1\mathtt{e}_n) \overset{\text{mult. by one}}{=} \mathtt{trace}_{n+1}(\mathtt{e}_n) \\
&\overset{(6.5)}{=} \mathtt{trace}_{n+1}(\mathtt{e}_{n-1}\mathtt{e}_n\mathtt{e}_{n-1}) \overset{\text{def. 6.5,ii)}}{=} \mathtt{trace}_{n+1}(\mathtt{e}_{n-1}\mathtt{e}_{n-1}\mathtt{e}_n) \\
&\overset{(6.3)}{=} \mathtt{trace}_{n+1}(q\mathtt{e}_{n-1}\mathtt{e}_n) \overset{\text{def. 6.5,i)}}{=} q\,\mathtt{trace}_{n+1}(\mathtt{e}_{n-1}\mathtt{e}_n) \\
&\overset{\text{def. 6.5,iii)}}{=} q\,\mathtt{trace}_n(\mathtt{e}_{n-1}) \overset{\text{def. 6.5,iii)}}{=} q\,\mathtt{trace}_{n-1}(1) \\
&\overset{\text{ind. assertion}}{=} qq^{n-1} = q^n.
\end{aligned}$$

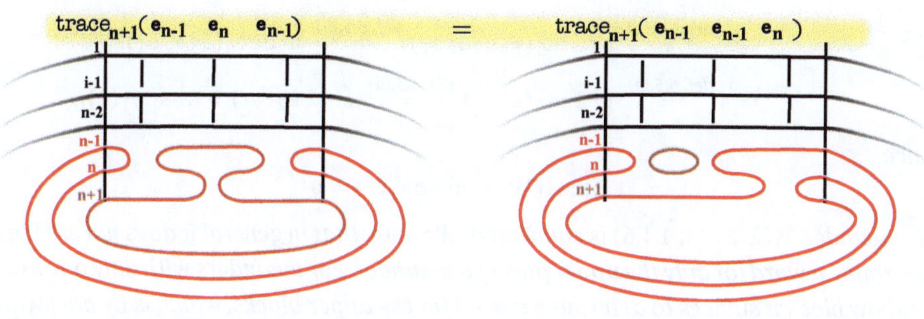

Fig. 6.8: Closure of strand diagrams for the fourth equality sign in the calculation (below) for $\mathtt{trace}_n(1) = q^n$.

By using the Markov trace, let us give some examples of how to determine connected components of a meander

$$\mathcal{M}(A_1, \ldots, A_N \mid B_1, \ldots, B_M)$$

in the Notation 1 with $2n = |\mathrm{V}^{\mathcal{M}}|$ the total number of vertices.

Example 6.1 (Evaluation of the Markov Trace). *We consider the following meander*

$$\mathcal{M}(2(2,1),1 \mid 6)$$

intersecting the straight line in $2n = 12$ intersection points.
By following Construction 17 we assign a word written in the generators of $\mathfrak{TL}_6(q)$ to the upper blocks given by $2(2,1),1$. As was explained before the words determined by using Dyck paths may differ since the generators commute on the same heights of the Dyck paths. Up to the order of the even or odd indexed generators, however, words are unique. Note, that it is also important to take the closure of the fully reduced words, because we may obtain a word[2] *which may be simplified by using the conditions of the Temperley–Lieb algebra on the generators. For $2(2,1),1$ we get, e.g., the following word*

$$\mathcal{W} = \mathtt{e}_4\mathtt{e}_3\mathtt{e}_5\mathtt{e}_2\mathtt{e}_4\mathtt{e}_1\mathtt{e}_3\mathtt{e}_4\mathtt{e}_5.$$

The closure of $\mathcal{W}$ is given by

$$\begin{aligned}
\mathtt{trace}_6(\mathcal{W}) &= \mathtt{trace}_6(\mathtt{e}_4\mathtt{e}_3\mathtt{e}_5\mathtt{e}_2\ \underbrace{\mathtt{e}_4\mathtt{e}_1}_{\mathtt{e}_1\mathtt{e}_4,\ (6.4)}\ \mathtt{e}_3\mathtt{e}_4\mathtt{e}_5)\\
&= \mathtt{trace}_6(\mathtt{e}_4\mathtt{e}_3\mathtt{e}_5\mathtt{e}_2\mathtt{e}_1\ \underbrace{\mathtt{e}_4\mathtt{e}_3\mathtt{e}_4}_{\mathtt{e}_4,\ (6.5)}\ \mathtt{e}_5)\\
&= \mathtt{trace}_6(\mathtt{e}_4\mathtt{e}_3\ \underbrace{\mathtt{e}_5(\mathtt{e}_2\mathtt{e}_1)}_{(\mathtt{e}_2\mathtt{e}_1)\mathtt{e}_5,\ def.\ 6.5,ii)}\ \mathtt{e}_4\mathtt{e}_5)\\
&= \mathtt{trace}_6(\mathtt{e}_4\mathtt{e}_3(\mathtt{e}_2\mathtt{e}_1)\ \underbrace{\mathtt{e}_5\mathtt{e}_4\mathtt{e}_5}_{\mathtt{e}_5,\ (6.5)})\\
&\overset{def.\ 6.5,iii)}{=} \mathtt{trace}_5(\mathtt{e}_4\mathtt{e}_3\mathtt{e}_2\mathtt{e}_1) \overset{def.\ 6.5,iii)}{=} \mathtt{trace}_4(\mathtt{e}_3\mathtt{e}_2\mathtt{e}_1)\\
&\overset{def.\ 6.5,iii)}{=} \mathtt{trace}_3(\mathtt{e}_2\mathtt{e}_1) \overset{def.\ 6.5,iii)}{=} \mathtt{trace}_2(\mathtt{e}_1) = \mathtt{trace}_1(1).
\end{aligned}$$

Since

$$\mathtt{trace}_6(\mathcal{W}) = \mathtt{trace}_1(1) = q^1$$

the meander $\mathcal{M}(2(2,1),1 \mid 6)$ is connected. We note, that in general it does not suffice to determine a word for only the upper part of a meander. For meanders with only one lower rainbow block it suffices to determine a word for the upper blocks, which is by backward-bottom-flip, see Construction 18, and its relation to the closure. Geometrically, taking the closure of a word means to close the upper and lower blocks following the rainbow rule as was mentioned before and illustrated in Figure 6.7.

---

2 Compare with Figure 6.6, in a).

6.4 Left and Right One-Shifts in Terms of Temperley–Lieb Algebra

The aim of this section is to take a look at Theorem 4.1 in terms of the Temperley–Lieb algebra. The examples which are discussed in this section also illustrate once again Construction 17 on building words from Dyck paths. Let

$$A_1, \ldots, A_n$$

be bracket expressions which represent general blocks. In this section we consider right one-shifts of closed meanders with same compositions of upper and lower blocks, i.e., Morse meanders of the form

$$\mathtt{shift}_{[1]}(\mathcal{M}(A_1, \ldots, A_n \mid A_1, \ldots, A_n)) =: \mathcal{M}_{[1]},$$

as was introduced in Chapter 4. Our aim is to study the associated words written with the aid of the generators of the Temperley–Lieb algebra. In order to proceed, we need to apply $\mathtt{coll}^{-1}$ to $\mathcal{M}_{[1]}$, since it is not closed. Then, we may determine the Dyck path of $\mathtt{coll}^{-1}(\mathcal{M}_{[1]})$. Let us consider an example. We consider the following right one-shifted meander

$$\mathtt{shift}_{[1]}(\mathcal{M}(1(2,1) \mid 1(2,1))) = \mathcal{M}^{coll}(2(4,2), 1 \mid 1, 2(4,2)),$$

see Figure 6.9. For $\mathtt{coll}^{-1}(\mathtt{shift}_{[1]}(\mathcal{M}(1(2,1) \mid 1(2,1))))$ we write down the following word expressed in terms of the generators of the Temperley–Lieb algebra $\mathfrak{TL}_{18}(q)$:

$$\mathcal{W}_{\mathcal{M}(2(4,2),1|1,2(4,2))} = \mathfrak{P}^+_{[1]}\mathfrak{I}^0_{18}\mathfrak{P}^-_{[1]}, \quad \text{where}$$

$$\begin{aligned}
\mathfrak{P}^+_{[1]} &= \overbrace{e_6}^{w_1}\overbrace{e_7e_5}^{w_2}\overbrace{e_4e_6e_8e_{12}}^{w_3}\overbrace{e_{13}e_{11}e_9e_7e_5e_3}^{w_4}\overbrace{e_2e_4e_6e_8e_{10}e_{12}e_{14}}^{w_5}\\
\mathfrak{I}^0_{18} &= e_1e_3e_5e_7e_9e_{11}e_{13}e_{15}e_{17}\\
\mathfrak{P}^-_{[1]} &= \underbrace{e_{16}e_{14}e_{12}e_{10}e_8e_6e_4}_{w'_5}\underbrace{e_5e_7e_9e_{11}e_{13}e_{15}}_{w'_4}\underbrace{e_{14}e_{10}e_8e_6}_{w'_3}\underbrace{e_9e_7}_{w'_2}\underbrace{e_8}_{w'_1}.
\end{aligned}$$

Note that $\mathfrak{P}^-_{[1]}$ is obtained by reversing the order and increasing all indices of the generators of $\mathfrak{P}^+_{[1]}$ by two. For $\mathtt{coll}^{-1}(\mathtt{shift}_{[-1]}(\mathcal{M}(1(1,2)|1(1,2))))$ we have:

$$\begin{aligned}
\mathfrak{P}^+_{[-1]} &= \overbrace{e_8}^{v_1}\overbrace{e_7e_9}^{v_2}\overbrace{e_6e_8e_{10}e_{14}}^{v_3}\overbrace{e_{15}e_{13}e_{11}e_9e_7e_5}^{v_4}\overbrace{e_4e_6e_8e_{10}e_{12}e_{14}e_{16}}^{v_5}\\
\mathfrak{I}^0_{18} &= e_1e_3e_5e_7e_9e_{11}e_{13}e_{15}e_{17}\\
\mathfrak{P}^-_{[-1]} &= \underbrace{e_{14}e_{12}e_{10}e_8e_6e_4e_2}_{v'_5}\underbrace{e_3e_5e_7e_9e_{11}e_{13}}_{v'_4}\underbrace{e_{12}e_8e_6e_4}_{v'_3}\underbrace{e_5e_7}_{v'_2}\underbrace{e_6}_{v'_1},
\end{aligned}$$

see the auxiliary Figure 6.11 and compare with the corresponding strand diagram in Figure 6.10. Hence, we obtain the following word

$$\mathcal{W}_{\mathcal{M}(1,2(4,2)|2(4,2)),1)} = \mathfrak{P}^+_{[-1]}\mathfrak{I}^0_{18}\mathfrak{P}^-_{[-1]}.$$

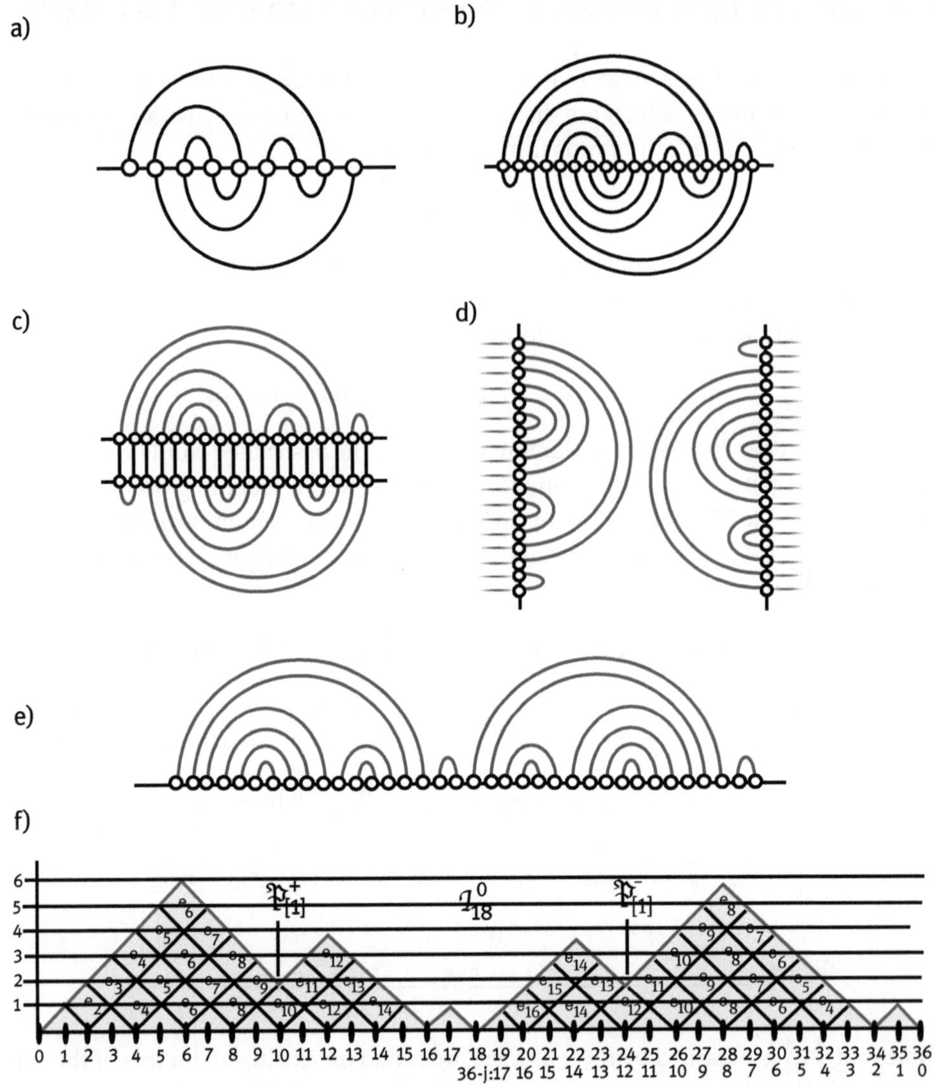

Fig. 6.9: a): right one-shifted meander $\texttt{shift}_{[1]}(\mathcal{M}(1(2,1) \mid 1(2,1))) = \mathcal{M}^{coll}(2(4,2),1 \mid 1,2(4,2))$; b): applying $\texttt{coll}^{-1}$ to $\mathcal{M}^{coll}(2(4,2),1 \mid 1,2(4,2))$; c),d): relating b) to the closure of its strand diagram; e) forward-bottom-flip of b); f) Dyck path corresponding to e).

By Theorem 4.1 we know that the Markov trace of both words is q^1 since it is associated to connected meanders. Without Theorem 4.1 we would end up with a lengthy calculation of the Markov trace.

In Figure 6.11, we give an alternative pictorial way of associating a word to

$$\texttt{shift}_{[-1]}(\mathcal{M}(1(1,2)|1(1,2))) \text{ and } \texttt{shift}_{[1]}(\mathcal{M}(1(1,2)|1(1,2)))$$

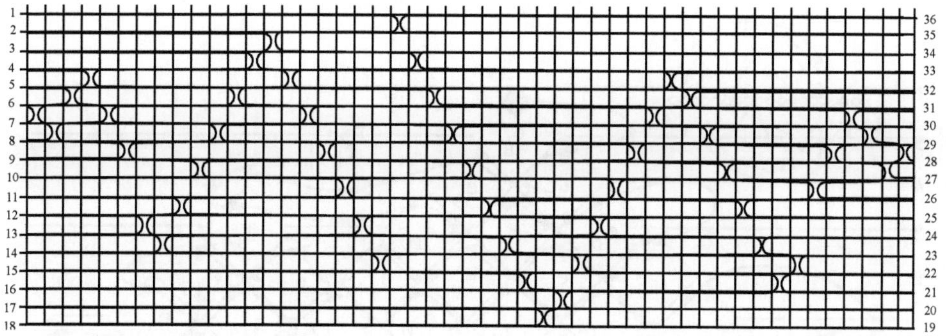

Fig. 6.10: Strand diagram determined from the Dyck path of meander $\mathcal{M}(2(4,2), 1 \mid 1, 2(4,2))$.

by depicting the Dyck path corresponding to lower blocks under the Dyck path corresponding to upper blocks. Obviously, both ways introduced in Figures 6.9 and 6.11 are equivalent by the backward/forward-bottom-flip of meanders, see Chapter 1.

In general, if we consider a right one-shifted meander

$$\mathcal{M}_{[1]} = \mathtt{shift}_{[1]}(\mathcal{M}(A_1, \dots, A_n \mid A_1, \dots, A_n))$$

having m intersection points with the straight line, we may assign a word $\mathcal{W}_{\mathcal{M}_{[1]}}$ in $\mathfrak{TL}_m(q)$ by considering closed meanders

$$\mathtt{coll}^{-1}(\mathcal{M}_{[1]}) = \mathtt{coll}^{-1}(\mathtt{shift}_{[1]}(\mathcal{M}(A_1, \dots, A_n \mid A_1, \dots, A_n))).$$

Let

$$\left(\cdot\right)_{\uparrow 2} : \mathsf{e}_i \cdot \ldots \cdot \mathsf{e}_k \mapsto \mathsf{e}_{i+2} \cdot \ldots \cdot \mathsf{e}_{k+2}$$

be the map raising all indices of generators of a word by two and similarly

$$\left(\cdot\right)_{\downarrow 2} : \mathsf{e}_i \cdot \ldots \cdot \mathsf{e}_k \mapsto \mathsf{e}_{i-2} \cdot \ldots \cdot \mathsf{e}_{k-2}$$

decreasing all indices of generators of a word by two. It is easy to see that in general the following correspondencies hold: $\mathfrak{P}^+_{[1]} = \left(\mathfrak{P}^+_{[-1]}\right)_{\downarrow 2}$ and $\mathfrak{P}^-_{[1]} = \left(\mathfrak{P}^-_{[-1]}\right)_{\uparrow 2}$. Moreover, by property (6.4): $\mathsf{e}_i\mathsf{e}_j = \mathsf{e}_j\mathsf{e}_i$ for $|i-j| > 1$, it holds:

$$\mathfrak{P}^+_{[1]} = \left(\mathfrak{P}^-_{[1]}\right)_{\uparrow 2} = \mathfrak{P}^-_{[-1]} \text{ and } \mathfrak{P}^+_{[-1]} = \left(\mathfrak{P}^-_{[-1]}\right)_{\downarrow 2} = \mathfrak{P}^-_{[1]}.$$

Therefore, we have the following relation between the underlying words $\mathcal{W}_{\mathtt{coll}^{-1}(\mathcal{M}_{[1]})}$ for right one-shifted elements and $\mathcal{W}_{\mathtt{coll}^{-1}(\mathcal{M}_{[-1]})}$ for left one-shifted elements:

$$\mathcal{W}_{\mathtt{coll}^{-1}(\mathcal{M}_{[1]})} = \mathfrak{P}^+_{[1]}\mathfrak{I}^0_m\mathfrak{P}^-_{[1]} = \mathfrak{P}^-_{[-1]}\mathfrak{I}^0_m\mathfrak{P}^+_{[-1]} = \mathtt{rev}(\mathcal{W}_{\mathtt{coll}^{-1}(\mathcal{M}_{[-1]})}),$$

where $\mathtt{rev}$ is the map which interchanges $\mathfrak{P}^+_{[\pm 1]}$ with $\mathfrak{P}^-_{[\pm 1]}$. The Markov trace evaluated in $\mathcal{W}_{\mathtt{coll}^{-1}(\mathcal{M}_{[\pm 1]})}$ yields the number of its connected components. By Theorem 4.1, $\mathcal{M}_{[1]}$ is connected and by Theorem 2.1 it follows that $\mathtt{coll}^{-1}(\mathcal{M}_{[1]})$ is also connected. Therefore, without explicit calculations in $\mathfrak{TL}_m(q)$, it holds: $\mathtt{trace}_m(\mathcal{W}_{\mathtt{coll}^{-1}(\mathcal{M}_{[\pm 1]})}) = q^1$.

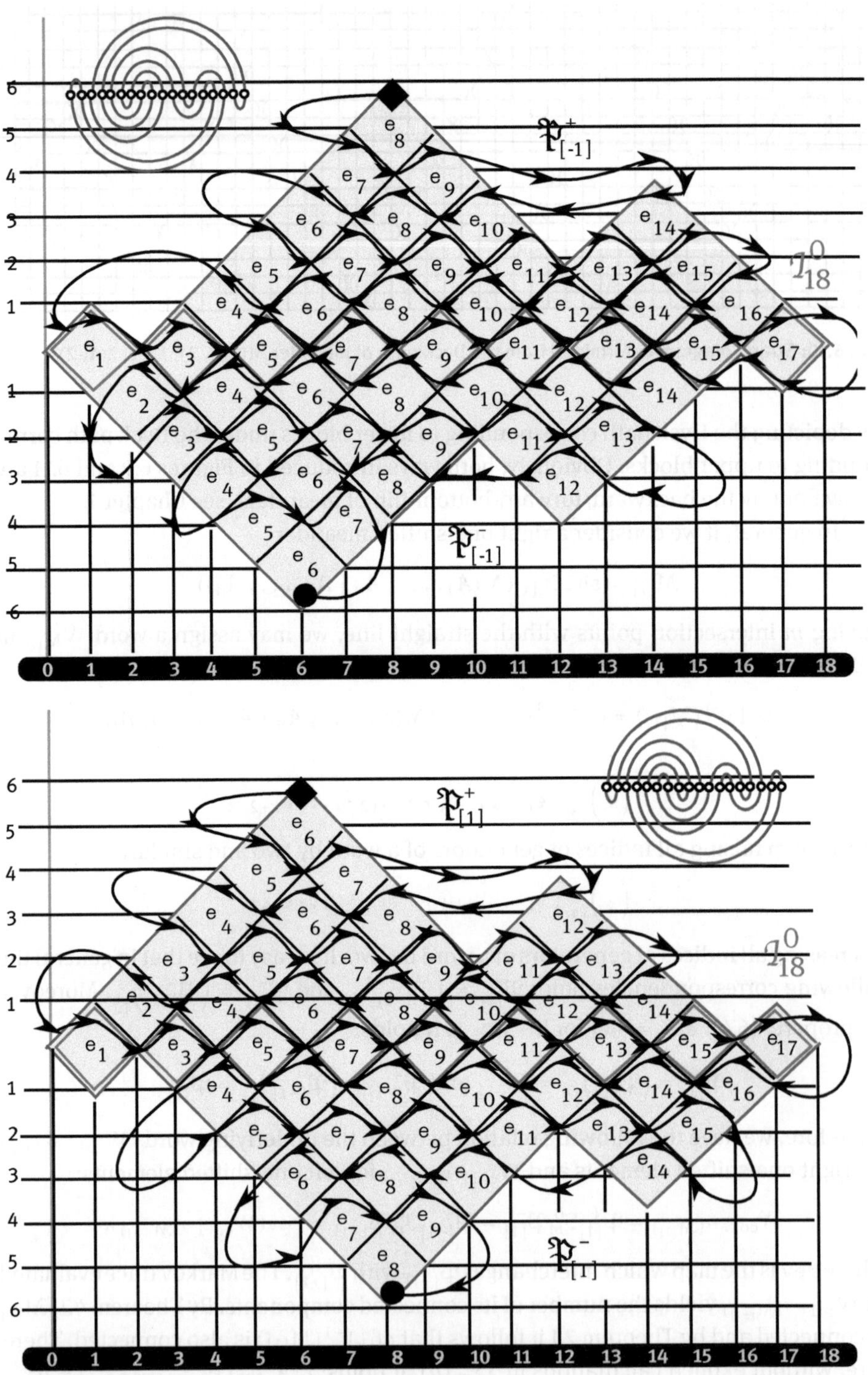

Fig. 6.11: Associating words to the left and right one-shifted meanders after applying the inverse of the collapse map, $\mathtt{coll}^{-1}(\mathtt{shift}_{[-1]}(\mathcal{M}(1(1,2)|1(1,2))))$ and $\mathtt{coll}^{-1}(\mathtt{shift}_{[1]}(\mathcal{M}(1(1,2)|1(1,2))))$.

7 Seaweed Lie Algebras and Seaweed Meanders

As was already mentioned in the Introduction, *seaweed meanders* are closely related to the so called *seaweed Lie algebras*. More precisely, the collapsed seaweed meanders, introduced as *meander graphs* in [7], are used for representing finite-dimensional Lie subalgebras of seaweed type of the general linear Lie algebra $\mathfrak{gl}_n$ (or special linear Lie algebra $\mathfrak{sl}_n$). The advantage of representing matrix Lie algebras of seaweed type by collapsed seaweed meanders, as was proposed in [7], lies in the simplification of the *index* computation. We therefore recall the definition of the *index* defined for Lie algebras. In the next chapter we will use the index in order to determine constant solutions of the classical Yang–Baxter equation by following [2].

7.1 Index of a Lie Algebra

Let $\mathfrak{g}$ be a finite-dimensional Lie algebra over some field $\mathbb{K}$, i.e., $\mathfrak{g}$ is a $\mathbb{K}$-vector space together with a Lie bracket operation

$$[\cdot,\cdot] : \mathfrak{g} \times \mathfrak{g} \to \mathfrak{g}$$

satisfying the following properties:

$$[A,[B,C]] + [B,[C,A]] + [C,[A,B]] = 0 \qquad \textit{(Jacobi identity)}$$
$$[A,B] = -[B,A], \qquad \textit{(skew-symmetry).}$$

Let $\mathfrak{g}^* := \mathtt{Hom}(V, \mathbb{K})$ denote the dual Lie algebra to $\mathfrak{g}$. Let

$$f : \mathfrak{g} \to \mathbb{K} \tag{7.1}$$

be a linear function, i.e., $f \in \mathfrak{g}^*$. With the aid of f, we introduce a bilinear skew-symmetric form

$$\begin{aligned} \mathcal{B}_f : \mathfrak{g} \times \mathfrak{g} &\to \mathbb{K}, \\ (A,B) &\mapsto f([A,B]) =: \mathcal{B}_f(A,B). \end{aligned} \tag{7.2}$$

Since the Lie bracket is bilinear and f linear, $\mathcal{B}_f$ is bilinear. By linearity of f and skew-symmetry of the Lie bracket

$$\mathcal{B}_f(A,B) = f([A,B]) = -f([B,A]) = -\mathcal{B}_f(B,A),$$

$\mathcal{B}_f$ is skew-symmetric. Since $\mathcal{B}_f$ is skew-symmetric the left and the right kernel of $\mathcal{B}_f$ coincide

$$\mathtt{Kern}(\mathcal{B}_f) := \{A \in \mathfrak{g} \mid \mathcal{B}_f(A,B) = 0 \ \forall B \in \mathfrak{g}\}. \tag{7.3}$$

DOI 10.1515/9783110533026-008

Definition 7.1 (Non-Degenerate). *A bilinear form $\mathcal{B}_f$ is called* non-degenerate *if its kernel is trivial, i.e.,*

$$\mathtt{Kern}(\mathcal{B}_f) := \{A \in \mathfrak{g} \mid \mathcal{B}_f(A, B) = 0 \; \forall B \in \mathfrak{g}\} = \{0\}.$$

We give the definition of the index of a Lie algebra $\mathfrak{g}$, cf. [7].

Definition 7.2. *Let $\mathfrak{g}$ be a finite-dimensional Lie algebra and $f \in \mathfrak{g}^*$. The index of $\mathfrak{g}$ is defined as the minimal dimension of the kernel of $\mathcal{B}_f$ as f ranges over $\mathfrak{g}^*$. In symbols:*

$$\mathtt{ind}(\mathfrak{g}) := \min_{f \in \mathfrak{g}^*} \mathtt{dim\,Kern}(\mathcal{B}_f) \tag{7.4}$$

The following example serves for an illustration, cf. [7].

Example 7.1. *(Index of the Lie algebra $\mathfrak{gl}_n$) We give the computation of the index of the Lie algebra $\mathfrak{gl}_n := \mathtt{Mat}(n, \mathbb{R})$. The trace*

$$\begin{aligned} \mathtt{Tr} : \mathtt{Mat}(n, \mathbb{R}) \times \mathtt{Mat}(n, \mathbb{R}) &\to \mathbb{R} \\ (A, B) &\mapsto \mathtt{Tr}(AB) \end{aligned}$$

is a bilinear form, which is symmetric, *i.e.,* $\mathtt{Tr}(AB) = \mathtt{Tr}(BA)$, non-degenerate, *i.e.,* $\mathtt{Kern}(\mathtt{Tr}) = \{0\}$. *For $A \in \mathtt{Mat}(n, \mathbb{R})$ we define $f_A(X) := \mathtt{Tr}(AX)$ and*

$$\mathcal{B}_{f_A}(X, Y) := (f_A[X, Y]).$$

Then, $\mathcal{B}_{f_A}$ is explicitly given by:

$$\mathcal{B}_{f_A}(X, Y) = \mathtt{Tr}(AXY - AYX) = \mathtt{Tr}(AXY - XAY) = \mathtt{Tr}([A, X], Y),$$

where we have used $\mathtt{Tr}(AB) = \mathtt{Tr}(BA)$ after the second equality sign. The last term in (7.1) shows that

$$X \in \mathtt{Kern}(\mathcal{B}_{f_A}) \quad \textit{if and only if} \quad [A, X] = 0.$$

When all eigenvalues of A are different, the space of all X which satisfies this condition is n-dimensional. All

$$f \in \mathtt{Hom}(\mathtt{Mat}(n, \mathbb{R}), \mathbb{R})$$

are of the form f_A due to the equality of the dimensions of $\mathtt{Mat}(n, \mathbb{R})^*$ and $\mathtt{Mat}(n, \mathbb{R})$. It follows that the index of $\mathfrak{gl}_n$ equals to n.

Definition 7.3. *We call a Lie algebra*

$$\mathfrak{g} \subset \mathfrak{sl}_n := \{A \in \mathtt{Mat}_n \mid \mathtt{Tr}(A) = 0\}$$

Frobenius, if ind($\mathfrak{g}$) $= 0$. For $\mathfrak{g} \subset \mathfrak{gl}_n$, we call $\mathfrak{g}$ Frobenius *if ind($\mathfrak{g}$) $= 1$. Frobenius Lie algebras will be denoted by $\mathfrak{f}$.*

7.2 Seaweed Lie Algebras

Seaweed Lie algebras were introduced by Kirillov and Dergachev in [7].

Definition 7.4. *Let* $\mathbb{K}$ *be an arbitrary field. Fix two ordered partitions* $\{a_i\}_{i=1}^r$ *and* $\{b_j\}_{j=1}^k$ *of the number n. Let* $\{e_m\}_{m=1}^n$ *be the standard basis in* $\mathbb{K}^n$*. A subalgebra of* $Mat(n, \mathbb{K})$ *that preserves the vector spaces*

$$V_i := span(e_1, \dots, e_{a_1+\dots+a_i}) \text{ and } W_j := span(e_{b_1+\dots+b_j+1}, \dots, e_n)$$

is called a subalgebra of seaweed type.

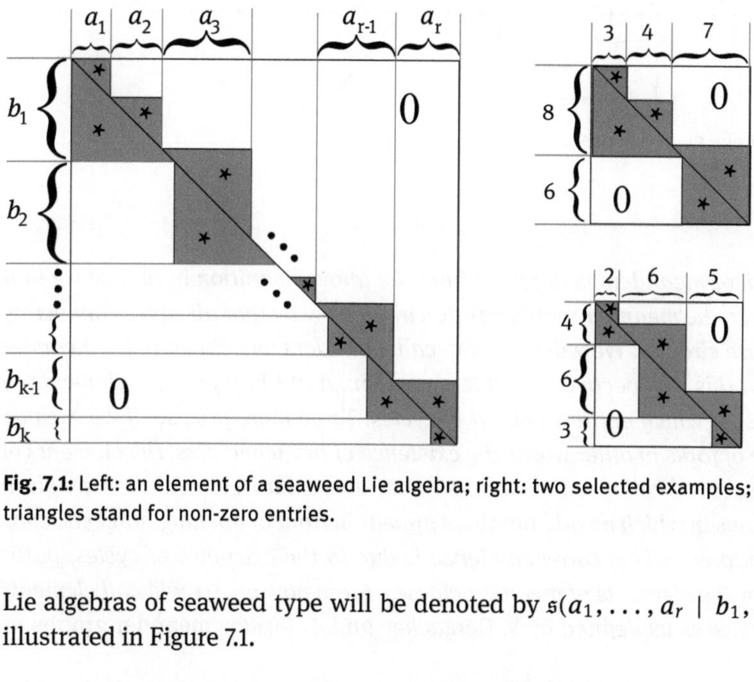

Fig. 7.1: Left: an element of a seaweed Lie algebra; right: two selected examples; stars in the shaded triangles stand for non-zero entries.

Lie algebras of seaweed type will be denoted by $\mathfrak{s}(a_1, \dots, a_r \mid b_1, \dots, b_k)$ and are illustrated in Figure 7.1.

7.3 Graph Representations of Seaweed Lie Algebras

We recall from [7] the definition of *meanders* by V. Dergachev and A. Kirillov. In the terminology of [7], segments are paths, circles are cycles and points are paths of length zero.

Definition 7.5 (Meander Representation, Meanders). *The* target space *T is* $\mathbb{R}^2$ *with an embedded line L. The* source space *M determines the type of the meander and is the disjoint union of circles, segments and points (one can also consider points as segments* *with 0 length). The* meander representation *is defined as an embedding of M into the target space T that satisfies the following conditions:*

i) *The endpoints of segments must map into L.*

ii) *Single points must map into L.*

iii) *The embedding is transversal to the line L.*

The meander *is defined as an equivalence class of meander representations under homeomorphisms of the target space. We give an example for elements of one meander class in Figure 7.2.*

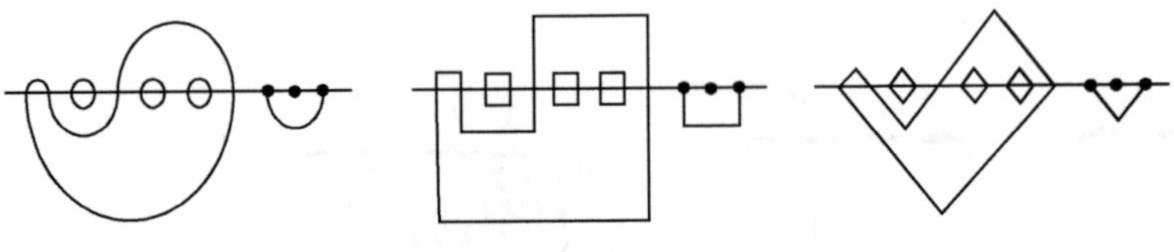

Fig. 7.2: Elements of one meander class.

Remarks 7.1.

i) *Our definition of meanders is different from the above definition in that sense that we have chosen the meander representation in such a way that all edges connecting the vertices are circular. We will therefore call meanders as defined in [7] meander classes. From this perspective, meanders as defined in Chapter 2 are elements of meander classes which entirely consist of cycles. To be more precise, if we exclude the existence of forks in other words the existence of branched arcs, the elements of meander classes are collapsed meanders whose non-collapsed meanders are given by compositions in which no odd numbers appear in front of opening brackets, compare with Chapter 2. This correspondence is due to the existence of cycles, paths and points for the elements of meander classes. From now on, we will call elements of meander classes as defined by V. Dergachev and A. Kirillov* meander graphs *as in [7].*

ii) *In the above definition, one might think that there is no explicit condition on how circles must be mapped under the meander representation. However, in the third condition of Definition 7.5 it is implicitly established. The embedding is transversal to the line L and therefore all exemplary cases shown in Figure 7.3, a), are excluded.*

Recall from [7] that to each meander graph, there is a permutation assigned in the following way: By labeling the intersection points $1, \ldots, n$ each arc in $\mathbb{E}^+$ and $\mathbb{E}^-$ of a meander graph defines a transposition $\tau = (j\ k)$ for $j, k \in \{1, \ldots, n\}$. The product of all transpositions over all arcs is the permutation associated to this meander. The permutation is not unique since the product can be taken in any order. However, the following theorem is valid, cf. [7]:

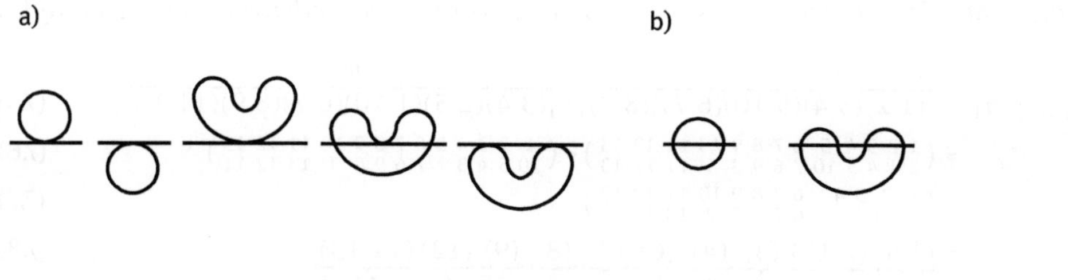

Fig. 7.3: a) Selected examples for how the meander representation shall not map circles; b) images of circles under the meander representation.

Theorem 7.1. *The number of cycles in a decomposition of the associated permutation is unique to the meander graph. It can be computed as twice the number of circles plus the number of segments plus the number of isolated points.*

The proof of the above theorem follows by induction, cf. [7].

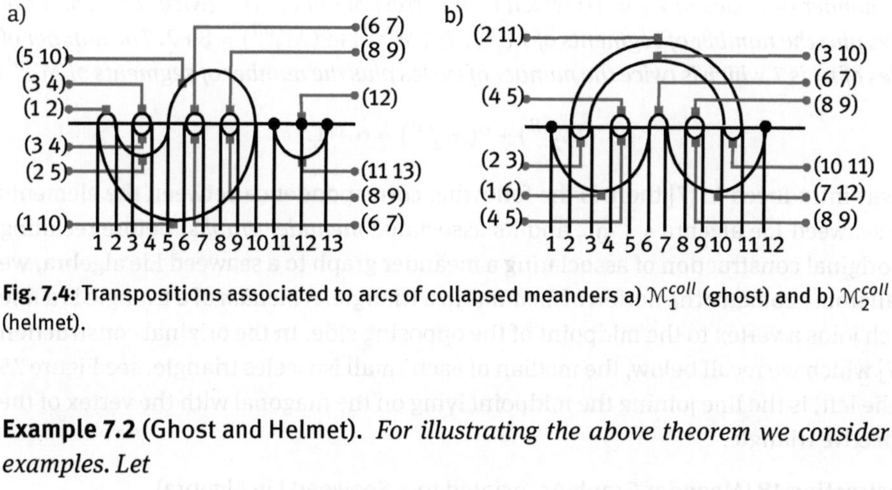

Fig. 7.4: Transpositions associated to arcs of collapsed meanders a) $\mathcal{M}_1^{coll}$ (ghost) and b) $\mathcal{M}_2^{coll}$ (helmet).

Example 7.2 (Ghost and Helmet). *For illustrating the above theorem we consider two examples. Let*

$$\mathcal{M}_1^{coll} := \mathcal{M}^{coll}(2, 2, 2(2, 2), 1, 1, 1 \mid 2(4, 2, 2), 3),$$
$$\mathcal{M}_2^{coll} := \mathcal{M}^{coll}(1, 4(2, 2, 2), 1 \mid 2(2, 2), 2(2, 2))$$

be two meander graphs or in other words two collapsed meanders without branched arcs, see Figure 7.4. We determine non-unique permutations by taking products (from the right) of the transpositions corresponding to arcs of the underlying meander graphs:

$$\begin{aligned}
\pi_1 &= \overbrace{[(1\ 2)(3\ 4)(5\ 10)(6\ 7)](8\ 9)]}^{in\ \mathbb{E}^+} \circ \overbrace{[(3\ 4)(2\ 5)(1\ 10)(6\ 7)(8\ 9)(11\ 13)]}^{in\ \mathbb{E}^-} && (7.5)\\
&= \begin{pmatrix} 1&2&3&4&5&6&7&8&9&10&11&12&13\\ 2&1&4&3&10&7&6&9&8&5&11&12&13 \end{pmatrix} \circ \begin{pmatrix} 1&2&3&4&5&6&7&8&9&10&11&12&13\\ 10&5&4&3&2&7&6&9&8&1&13&12&11 \end{pmatrix} && (7.6)\\
&= \begin{pmatrix} 1&2&3&4&5&6&7&8&9&10&11&12&13\\ 5&10&3&4&1&6&7&8&9&2&13&12&11 \end{pmatrix} && (7.7)\\
&= \underbrace{(1\ 5)}_{1}\underbrace{(2\ 10)}_{2}\underbrace{(3)}_{3}\underbrace{(4)}_{4}\underbrace{(6)}_{5}\underbrace{(7)}_{6}\underbrace{(8)}_{7}\underbrace{(9)}_{8}\underbrace{(12)}_{9}\underbrace{(11\ 13)}_{10} && (7.8)\\
\pi_2 &= [(2\ 11)(4\ 5)(3\ 10)(6\ 7)(8\ 9)] \circ [(2\ 3)(1\ 6)(4\ 5)(10\ 11)(7\ 12)(8\ 9)]\\
&= \begin{pmatrix} 1&2&3&4&5&6&7&8&9&10&11&12\\ 1&11&10&5&4&7&6&9&8&3&2&12 \end{pmatrix} \circ \begin{pmatrix} 1&2&3&4&5&6&7&8&9&10&11&12\\ 6&3&2&5&4&1&12&9&8&11&10&7 \end{pmatrix} && (7.9)\\
&= \begin{pmatrix} 1&2&3&4&5&6&7&8&9&10&11&12\\ 7&10&11&4&5&1&12&8&9&2&3&6 \end{pmatrix} && (7.10)\\
&= \underbrace{(1\ 7\ 12\ 6)}_{1}\underbrace{(2\ 10)}_{2}\underbrace{(3\ 11)}_{3}\underbrace{(4)}_{4}\underbrace{(5)}_{5}\underbrace{(8)}_{6}\underbrace{(9)}_{7}, && (7.11)
\end{aligned}$$

where π_j corresponds to

$$\mathcal{M}_j^{coll}, \quad j = 1, 2.$$

The number of cycles of π_1 is 10 *which is in correspondence with twice the number of cycles plus the number of segments of $\mathcal{M}_1^{coll}$: $2\mathrm{C}(\mathcal{M}_1^{coll}) + \mathrm{P}(\mathcal{M}_1^{coll}) = 8+2$. The number of cycles of π_2 is 7 which is twice the number of cycles plus the number of segments of $\mathcal{M}_2^{coll}$:*

$$2\mathrm{C}(\mathcal{M}_2^{coll}) + \mathrm{P}(\mathcal{M}_2^{coll}) = 6 + 1.$$

As was introduced in [7] there is the following correspondence between the elements of a seaweed Lie algebra $\mathfrak{s} \subseteq \mathfrak{gl}_n$ and its associated *meander graphs*. Before recalling the original construction of associating a meander graph to a seaweed Lie algebra, we recall from geometry the notion of *medians* in a triangle. A *median* of a triangle is a line which joins a vertex to the midpoint of the opposing side. In the original construction of [7] which we recall below, the median of each small isosceles triangle, see Figure 7.5 on the left, is the line joining the midpoint lying on the diagonal with the vertex of the respective triangle.

Construction 18 (Meander Graph Associated to a Seaweed Lie Algebra).

Step 1. *We distinguish special entries in the matrix being an element of a given seaweed algebra by coloring the entries black, gray, red and white. The diagonal non-zero entries are colored gray, the non-zero entries lying on the medians of small triangles are colored red, the remaining non-zero entries are colored black and all remaining entries being zero are colored white.*

Step 2. *To each seaweed algebra there is a* meander graph *associated to it in the following way: Connect each red entry with the gray entry in the same column or row in the matrix.*

Note that by connecting each red entry with the gray entry in the same column or row in the matrix we obtain a rainbow lower or upper block up to homeomorphisms, see

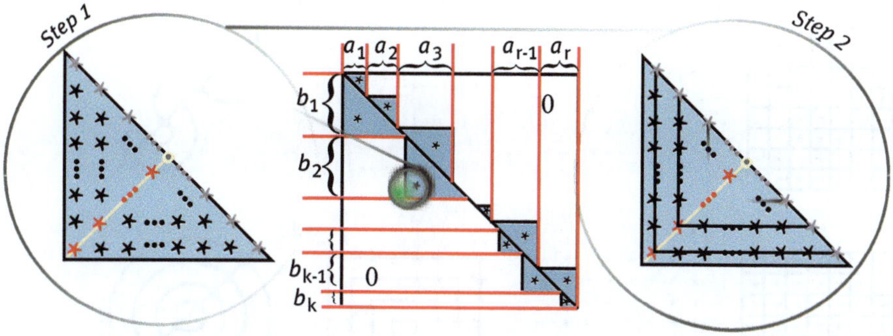

Fig. 7.5: Illustration of Step 1 and Step 2 of the above construction; left: median is highlighted in yellow; right: each red and gray entry are connected by straight lines.

Figure 7.5. Therefore, by assuming the connections to be half-circles, it is easy to see, that meander graphs associated to elements of seaweed Lie algebras are equivalent to collapsed seaweed meanders as was defined in Chapter 1. Let us illustrate the above Construction 18 via the following examples:

Examples 7.1 (Collapsed Seaweed Meander Associated to an Element of $\mathfrak{s}$). *Consult Figure 7.6 when reading the examples.*

1) *Let $n = 8$ and*

$$\mathfrak{s}(3, 3, 2 \mid 4, 4)$$

be a seaweed Lie algebra, i.e., elements of $\mathfrak{s}(3, 3, 2 \mid 4, 4)$ are of the form as depicted in a). We color the entries of the above matrix according to Step 1 of the Construction 18 in b). By connecting the red entries with the gray entries we obtain the meander graph in c), finishing Step 2 of our example. In d) we depict the associated collapsed seaweed meander

$$\mathcal{S}^{coll}(3, 3, 2 \mid 4, 4).$$

2) *The matrices of the general linear Lie algebra $\mathfrak{gl}_n$, see a'), have gray colored diagonal and red colored anti-diagonal entries in b'). By connecting them analogously to 1) we end up with a meander graph in c'). The resulting meander graph consists of four squares which is equivalent to a collapsed seaweed meander $\mathcal{S}^{coll}(8 \mid 8)$ in d').*

Let us now recall the following *Main theorem* of [7].

Theorem 7.2. *The index of a seaweed algebra $\mathfrak{s} \subset \mathfrak{gl}_n$ is equal to:*

i) *twice the number of cycles in the associated meander graph plus the number of segments plus the number of isolated points,*

ii) *the number of cycles in the permutation corresponding to the associated meander graph.*

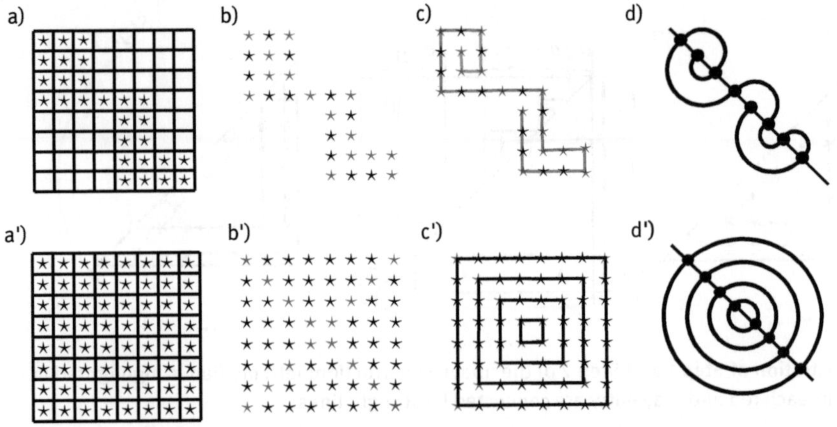

Fig. 7.6: a)–d): collapsed seaweed meander $\mathcal{S}^{coll}(3,3,2 \mid 4,4)$ (in d)) associated to matrices (in a)) of seaweed Lie algebra $\mathfrak{s}(3,3,2 \mid 4,4)$; a')–d'): collapsed seaweed meander $\mathcal{S}^{coll}(8 \mid 8)$ (in d)) associated to matrices (in a')) of seaweed Lie algebra $\mathfrak{s}(8 \mid 8)$; b),b') and c),c') are the two steps described in Construction 18.

The proof of the above theorem may be found in the original work [7] in detail. By combining Theorem 1.1 with the above Theorem 7.2, we see a more direct way to assign a seaweed meander to a seaweed Lie algebra rather than its collapsed form as proposed in [7]. The number of connected components of a seaweed meander equals the number of Jordan curves. Instead of following the Construction 18, as was presented in [7] in order to associate a collapsed seaweed meander to a seaweed Lie algebra, we present an alternative way, here. Namely, we associate a seaweed meander to a seaweed Lie algebra described in the construction given below.

Construction 19 (Seaweed Meander Associated to a Seaweed Lie Algebra). *The non-zero entries in the matrix of a given seaweed algebra $\mathfrak{s}(a_1, \dots, a_N \mid b_1, \dots, b_M)$ define two path diagrams P^+ and P^-. The path diagram P^+ is a graph in $\mathbb{R}^2$ on the vertices*

$$v_i = (i, h_i^+) \text{ for } 0 \le i \le 2n \text{ such that } v_j = (j, a_j) \text{ for } j = \left\lfloor \frac{a_j}{2} \right\rfloor .$$

Similarly, define P^-. Recall, from the definition of a path diagram 6.4, that h_i^+ is the number of arcs which pass over the midperpendicular between the ith and $(i+1)^{st}$ vertex of the upper block $\mathcal{B}^+$ represented by $a_1, \dots, a_N$. This correspondence holds between the path diagram P^- and the lower block $\mathcal{B}^-$ given by $b_1, \dots, b_M$ in the analogous way. In this way we obtain a seaweed meander

$$\mathcal{S}(a_1, \dots, a_N \mid b_1, \dots, b_M)$$

associated to a seaweed Lie algebra $\mathfrak{s}(a_1, \dots, a_N \mid b_1, \dots, b_M)$.

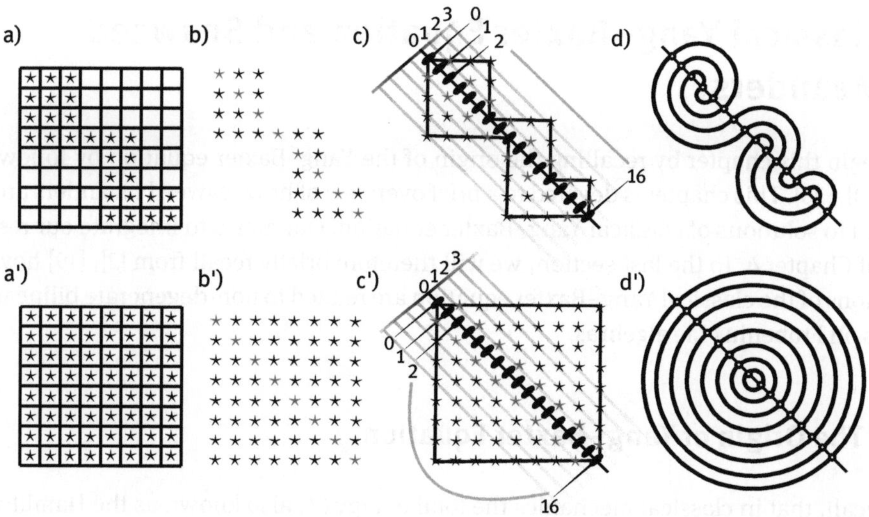

Fig. 7.7: Seaweed meander associated to a seaweed Lie algebra illustrated on two examples; In a), a'): matrix of a seaweed Lie algebra with non-zero entries indicated by a " $\star$ " and all other entries being zero; b),b'): colored diagonal; c),c'): construction of Dyck path by duplicating the diagonal; d), d'): seaweed meander obtained from the corresponding Dyck paths.

Note, that the non-zero entries on the diagonal in the matrix of a given seaweed Lie algebra correspond to the first heights of the path diagrams P^+, respectively P^-.

Let us illustrate the above construction by considering two examples in Figure 7.7. For this, let $\mathfrak{s}(3,3,2 \mid 4,4)$ and $\mathfrak{s}(8 \mid 8)$ be two seaweed Lie algebras whose elements are illustrated in a) and a'), Figure 7.7. By following Construction 19 we associate seaweed meander $\mathcal{S}(3,3,2 \mid 4,4)$ to $\mathfrak{s}(3,3,2 \mid 4,4) \subset \mathfrak{gl}_8$ and $\mathcal{S}(8 \mid 8)$ to $\mathfrak{s}(8 \mid 8) = \mathfrak{gl}_8$. By Theorems 7.2 and 1.1, $\mathfrak{s}(3,3,2 \mid 4,4)$ is *Frobenius*, whereas $\mathfrak{s}(8 \mid 8)$ is not, since in the first case the associated seaweed meander is connected and in the latter not.

Given two compositions – $(a_1,\ldots,a_n)$ representing n rainbow blocks in $\mathbb{E}^+$ and $(b_1,\ldots,b_n)$ in $\mathbb{E}^-$ – it is not clear whithout applying algorithms such as in [18], whether a seaweed meander is in the set of non-collapsed Frobenius meanders. We finish this chapter by noting that our results of Chapter 3 on Morse meanders obtained by applying the right one-shift map $\mathtt{shift}_{[1]}$ to seaweed meanders of the form $\mathcal{S}(a_1,\ldots,a_n \mid a_1,\ldots,a_n)$ are, apart from being Morse, collapsed *Frobenius* meanders, i.e., they represent seaweed Lie algebras $\mathfrak{f} \subset \mathfrak{gl}_n$ of index one. Recall that seaweed Lie algebras of index one are called Frobenius Lie algebras. The aim of the next chapter consists in recalling the relation of Frobenius Lie algebras and constant solutions of the classical Yang–Baxter equation by following [2].

8 Classical Yang–Baxter Equation and Seaweed Meanders

We begin this chapter by recalling the origin of the Yang–Baxter equation by following [30], [26]. This chapter is devoted to a brief overview of how seaweed meanders are related to solutions of classical Yang–Baxter equation. Our aim is to integrate our results of Chapter 4. In the last section, we will therefore briefly recall from [2], [19] how solutions of the classical Yang–Baxter equation are related to non-degenerate bilinear forms on Frobenius Lie algebras.

8.1 The Origin of Yang–Baxter Equation

We recall, that in classical mechanics the total energy $\mathcal{H}$, also known as the Hamiltonian, of a particle is given by the sum of kinetic energy $\mathcal{K}$ and potential energy V. In non-relativistic quantum mechanics, the Hamiltonian for a particle with mass m in one dimension, reads as follows:

$$\mathrm{H}_1 = \frac{p^2}{2m} + V(x), \quad p = -i\hbar\frac{\partial}{\partial x}. \tag{8.1}$$

Note that plugging p into H_1 yields

$$\mathrm{H}_1 = -\frac{\hbar^2}{2m}\frac{\partial^2}{\partial x^2} + V(x). \tag{8.2}$$

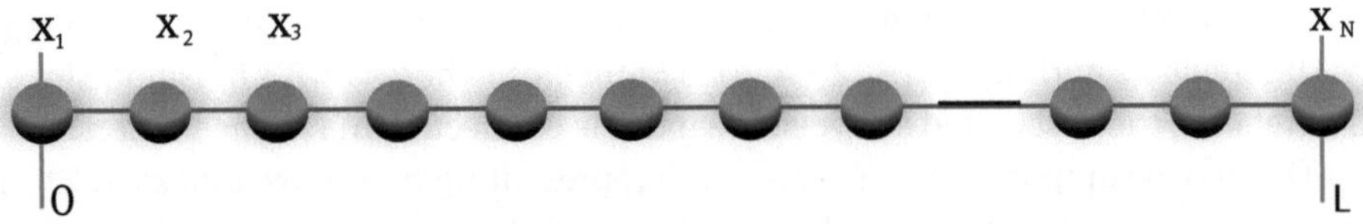

Fig. 8.1: Particles ordered on a line. The interaction of neighboring particles is marked by green regions.

Originally, the Yang–Baxter equation was derived in the reduction process of the following problem, cf. [30]: Consider N particles $x_1, \ldots, x_N$ with $m = \hbar = 1$ ordered on a line of length L, see Figure 8.1. The repulsive δ-function interaction of the particles is modeled by the following Hamiltonian operator on a Hilbert space X:

$$\mathrm{H}_N : X \to X \tag{8.3}$$

$$\psi \mapsto -\sum_{i=1}^{N} \frac{\partial^2}{\partial x_i^2}\psi + 2c\sum_{i<j}\delta(x_i - x_j)\psi. \tag{8.4}$$

DOI 10.1515/9783110533026-009

Remarks 8.1.

1. *The function ψ in (8.3) is called the wave function.*
2. *If $c < 0$, where c is the coupling constant, then the interaction of the particles is called attractive, if $c = 0$ it is called free, and if $c > 0$, repulsive. Hence, in (8.4) $c > 0$.*
3. *Physicists define the Dirac "δ -function" as:*

$$\delta(x_a - x_b) = \begin{cases} \infty, & \textit{for} \quad x_a = x_b \\ 0, & \textit{otherwise} \end{cases} \tag{8.5}$$

$$\int_{-\infty}^{+\infty} \delta(x)dx = 1. \tag{8.6}$$

For a rigorous mathematical definition of the Dirac δ-distribution we refer the reader to [31, p. 1047].

Since the Hamiltonian operator does not depend on time, the energy spectrum of the system is determined by the *time-independent Schrödinger equation*:

$$\mathrm{H}_N\, \psi = E\psi, \tag{8.7}$$

in which, different to the *time-dependent Schrödinger equation*, the imaginary unit *i* is not presented in front of $E\psi$. As the particles interact on a line L, the problem is called one-dimensional. Assume, without loss of generality, that the particles are ordered in the following way:

$$0 \leq x_1 < x_2 < x_3 < \cdots < x_N \leq L. \tag{8.8}$$

Let $\pi, \sigma \in S_N$ be two permutations, i.e.,

$$\pi = \begin{pmatrix} 1 & 2 & 3 & \dots & N \\ \pi(1) & \pi(2) & \pi(3) & \dots & \pi(N) \end{pmatrix} \text{ and } \sigma = \begin{pmatrix} 1 & 2 & 3 & \dots & N \\ \sigma(1) & \sigma(2) & \sigma(3) & \dots & \sigma(N) \end{pmatrix}.$$

Using Bethe's Ansatz, the wave function can be expanded as follows:

$$\psi(x_1, \dots, x_N) = \sum_{\sigma \in S_N} \theta(x_\sigma) \sum_{\pi \in S_N} A_\pi^\sigma exp(i\, \kappa_\pi\, x_\sigma), \tag{8.9}$$

where $\theta(x_\sigma) = \theta(x_{\sigma(1)}, \dots, x_{\sigma(N)}) = \begin{cases} 1 & \text{if } 0 \leq x_{\sigma(1)} < \cdots < x_{\sigma(N)} \leq L \\ 0 & \text{otherwise.} \end{cases}$

Bethe's Ansatz assumes that two particles which collide with each other have to be neighboring each other with positions, denoted by $x_{\sigma(j)}$ and $x_{\sigma(j+1)}$. Without loss of generality, assume that $x_{\sigma(j)} < x_{\sigma(j+1)}$ before and $x_{\sigma(j+1)} < x_{\sigma(j)}$ after the two particles with the positions $x_{\sigma(j)}$ and $x_{\sigma(j+1)}$ collide.

Define $\sigma' := \sigma(j\ j+1)$, where $(j\ j+1)$ denotes the transposition. In the following we derive the Yang–Baxter equation which first appeared in [30]. This will be achieved from the continuity condition of the wave function ψ and the proper discontinuity of its first derivative, which are given by:

$$\forall\, j \in \{1, \ldots, N-1\} : \lim_{h \searrow 0} \psi|_{x_{\pi(j)}=x_{\pi(j+1)}+h} = \lim_{h \nearrow 0} \psi|_{x_{\pi(j)}=x_{\pi(j+1)}+h}, \tag{8.10}$$

$$\lim_{h\searrow 0}\left(\frac{\partial\psi}{\partial x_{\pi(j)}} - \frac{\partial\psi}{\partial x_{\pi(j+1)}}\right) - \lim_{h\nearrow 0}\left(\frac{\partial\psi}{\partial x_{\pi(j)}} - \frac{\partial\psi}{\partial x_{\pi(j+1)}}\right) = 2c\psi|_{x_{\pi(j)}=x_{\pi(j+1)}}. \tag{8.11}$$

Equation (8.11) is obtained as follows: By plugging (8.3) into (8.7) and then integrating from $-\epsilon$ to $+\epsilon$ we obtain:

$$-\int_{-\epsilon}^{+\epsilon} \psi''(x)dx + 2c\int_{-\epsilon}^{+\epsilon} \delta(x)\psi(x)dx = E\int_{-\epsilon}^{+\epsilon} \psi(x)dx. \tag{8.12}$$

The right hand sight of (8.12) becomes zero for ϵ tending to zero since ψ is continuous. By the property of the δ-distribution, it holds:

$$\lim_{\epsilon\to 0}\int_{-\epsilon}^{+\epsilon} \delta(x)\psi(x)dx = \psi(0).$$

Therefore,

$$\int_{-\epsilon}^{+\epsilon} \psi''(x)dx = \psi'(-\epsilon) - \psi'(+\epsilon) = 2c\psi(0).$$

Hence, for the collision of two particles with the positions $x_{\pi(i)}$ and $x_{\pi(i+1)}$ we end up with (8.11). The above equations (8.10) and (8.11) imply the following conditions on the amplitudes:

$$\forall j \in \{1, \ldots, N-1\} : \qquad A^{\sigma}_{\pi} + A^{\sigma}_{\pi'} = A^{\sigma'}_{\pi} + A^{\sigma'}_{\pi'}, \tag{8.13}$$

$$i(\kappa_{\pi(j)} - \kappa_{\pi(j+1)})(A^{\sigma'}_{\pi'} - A^{\sigma'}_{\pi} + A^{\sigma}_{\pi'} - A^{\sigma}_{\pi}) = 2c(A^{\sigma}_{\pi} - A^{\sigma}_{\pi'}) \tag{8.14}$$

Plugging (8.13) into (8.14) yields:

$$A^{\sigma}_{\pi'} = \frac{cA^{\sigma}_{\pi}}{i(\kappa_{\pi(j)} - \kappa_{\pi(j+1)}) - c} + \frac{i(\kappa_{\pi(j)} - \kappa_{\pi(j+1)})A^{\sigma'}_{\pi}}{i(\kappa_{\pi(j)} - \kappa_{\pi(j+1)}) - c}. \tag{8.15}$$

Let ξ_{π} and $\xi_{\pi'}$ denote the column vectors of the following matrix[1]

$$\Upsilon = \begin{pmatrix} A^{\sigma_1}_{\pi_1} & A^{\sigma_1}_{\pi_2} & \ldots & A^{\sigma_1}_{\pi_{N!}} \\ A^{\sigma_2}_{\pi_1} & A^{\sigma_2}_{\pi_2} & \ldots & A^{\sigma_2}_{\pi_{N!}} \\ \vdots & \vdots & \ldots & \vdots \\ A^{\sigma_{N!}}_{\pi_1} & A^{\sigma_{N!}}_{\pi_2} & \ldots & A^{\sigma_{N!}}_{\pi_{N!}} \end{pmatrix}.$$

1 For obtaining $\xi_{\pi'}$ interchange π with π' in Υ.

Then, the equation (8.15) may be written in the following form, cf. [30]:

$$\xi_{\pi'} = Y^{j,j+1}_{\pi(j),\pi(j+1)}\xi_{\pi}, \tag{8.16}$$

where $Y^{i,j}_{a,b}$ denote the operators defined in the following way:

$$Y^{a,b}_{k,l} := -\frac{x_{k,l}}{1+x_{k,l}} + \frac{1}{1+x_{k,l}}\hat{P}_{a,b}, \tag{8.17}$$

$$x_{k,l} := \frac{ic}{\kappa_k - \kappa_l}, \tag{8.18}$$

where $\hat{P}_{a,b}$ is permuting $\sigma(a)$ with $\sigma(b)$ in the coefficients of the lth-column vector of Y, ξ_{π_l}, i.e.,

$$\hat{P}_{a,b} : \mathtt{Mat}(N! \times 1, \mathbb{R}) \to \mathtt{Mat}(N! \times 1, \mathbb{R}) \tag{8.19}$$

$$\begin{pmatrix} \vdots \\ A^{\sigma(a)}_{\pi_l} \\ \vdots \\ A^{\sigma(b)}_{\pi_l} \\ \vdots \end{pmatrix} \mapsto \begin{pmatrix} \vdots \\ A^{\sigma(b)}_{\pi_l} \\ \vdots \\ A^{\sigma(a)}_{\pi_l} \\ \vdots \end{pmatrix}. \tag{8.20}$$

In [30], the author claims the following fact which reflects the meaning of the Yang–Baxter equation and which can be formulated as

Theorem 8.1. *Bethe's Ansatz, as introduced in* (8.9), *on the wave function ψ of the boundary value problem defined by the equation*

$$H\psi = E\psi$$

with boundary conditions (8.10) *and* (8.11) *is consistent if and only if:*

$$Y^{a,b}_{i,j} \cdot Y^{a,b}_{j,i} = \mathtt{Id} \tag{8.21}$$

$$Y^{a,b}_{j,k} \cdot Y^{b,c}_{i,k} \cdot Y^{b,c}_{i,j} = Y^{b,c}_{i,j} \cdot Y^{a,b}_{i,k} \cdot Y^{b,c}_{j,k}. \tag{8.22}$$

Equation (8.22) is called the Yang–Baxter equation (YBE) and can be rewritten in the following form:

$$\mathrm{R}_{12}(\lambda_1 - \lambda_2)\mathrm{R}_{13}(\lambda_1 - \lambda_3)\mathrm{R}_{23}(\lambda_2 - \lambda_3) = \mathrm{R}_{23}(\lambda_2 - \lambda_3)\mathrm{R}_{13}(\lambda_1 - \lambda_3)\mathrm{R}_{12}(\lambda_1 - \lambda_2). \tag{8.23}$$

8.2 YBE and the Braid Relation

The aim of this section is to show how we can derive the Artin's braid group relation from Yang–Baxter equation.

Let V be a finite-dimensional vector space over $\mathbb{K}$, $\mathtt{dim}(V) = n$. Let

$$\mathrm{R} \in \mathtt{End}(V \otimes V) \text{ and } P \in \mathtt{End}(V \otimes V)$$

be the permutation endomorphisms, i.e., for $a, b \in V$:

$$P(a \otimes b) = b \otimes a \text{ and } \mathtt{Id} \in \mathtt{End}(V)$$

the identity endomorphism, then define:

$$\mathrm{R}_{12} := \mathrm{R} \otimes \mathtt{Id}, \qquad \mathrm{R}_{23} := \mathtt{Id} \otimes \mathrm{R}, \tag{8.24}$$

$$\mathrm{R}_{13} := P_{23}\mathrm{R}_{12}P_{23} = (\mathtt{Id} \otimes P)(\mathrm{R} \otimes \mathtt{Id})(\mathtt{Id} \otimes P). \tag{8.25}$$

Note that after fixing a basis in V, we may express R by a $n^2 \times n^2$-matrix, $\mathtt{Id}$ becomes a $n \times n$-identity matrix and for, e.g., $n = 2$:

$$P = \begin{pmatrix} 1 & 0 & 0 & 0 \\ 0 & 0 & 1 & 0 \\ 0 & 1 & 0 & 0 \\ 0 & 0 & 0 & 1 \end{pmatrix}$$

is a permutation matrix. Then with the aid of (8.24), (8.25), the parameter dependent Yang–Baxter equation is written in the following form:

$$\mathrm{R}_{12}(\lambda_1 - \lambda_2)\mathrm{R}_{13}(\lambda_1 - \lambda_3)\mathrm{R}_{23}(\lambda_2 - \lambda_3) = \mathrm{R}_{23}(\lambda_2 - \lambda_3)\mathrm{R}_{13}(\lambda_1 - \lambda_3)\mathrm{R}_{12}(\lambda_1 - \lambda_2), \tag{8.26}$$

where $\lambda_i \in \mathbb{C}$ are the spectral parameters corresponding to the vectors of the wave function ψ. We recall the following proposition proven in [27].

Proposition 8.1. *Let* $\mathrm{R} \in \mathtt{End}(V \otimes V)$ *be a solution to the (YBE), that is:*

$$\mathrm{R}_{12}(\mu)\mathrm{R}_{13}(\lambda_1)\mathrm{R}_{23}(\lambda_2) = \mathrm{R}_{23}(\lambda_2)\mathrm{R}_{13}(\lambda_1)\mathrm{R}_{12}(\mu).$$

Then $\hat{\mathrm{R}} := P\mathrm{R}$ *is a solution to*

$$\hat{\mathrm{R}}_{23}(\lambda_2)\hat{\mathrm{R}}_{12}(\lambda_1)\hat{\mathrm{R}}_{23}(\mu) = \hat{\mathrm{R}}_{12}(\mu)\hat{\mathrm{R}}_{23}(\lambda_1)\hat{\mathrm{R}}_{12}(\lambda_2).$$

Proof. We set $\mu := \lambda_1 - \lambda_2$ and $\lambda_3 = 0$ for simplicity. Then, the (YBE) in (8.26) becomes:

$$\mathrm{R}_{12}(\mu)\mathrm{R}_{13}(\lambda_1)\mathrm{R}_{23}(\lambda_2) = \mathrm{R}_{23}(\lambda_2)\mathrm{R}_{13}(\lambda_1)\mathrm{R}_{12}(\mu). \tag{8.27}$$

Let P_{ij} be the permutation on the ith and j^{th} component of $\bigotimes^{\infty} V$ and define

$$\hat{\mathrm{R}}_{ij} = P_{ij}R_{ij}.$$

Then, consider

$$\begin{aligned}\hat{R}_{23}(\lambda_2)\hat{R}_{12}(\lambda_1)\hat{R}_{23}(\mu) &= (P_{13}\underbrace{P_{12}\hat{R}_{12}(\lambda_2)}_{R_{12}(\lambda_2)}\underbrace{P_{13}\underbrace{P_{23})(P_{23}}_{\mathrm{Id}}\hat{R}_{13}(\lambda_1)P_{23})}_{R_{13}(\lambda_1)}\underbrace{P_{23})\hat{R}_{23}(\mu)}_{R_{23}(\mu)}\\ &= P_{13}R_{12}(\lambda_2)R_{13}(\lambda_1)R_{23}(\mu)\\ &= P_{13}\cdot(\text{left hand side of the (YBE)}).\\ \hat{R}_{12}(\mu)\hat{R}_{23}(\lambda_1)\hat{R}_{12}(\lambda_2) &= (P_{13}\underbrace{P_{23}\hat{R}_{23}(\mu)}_{R_{23}(\mu)}\underbrace{P_{13}\underbrace{P_{12})(P_{12}}_{\mathrm{Id}}\hat{R}_{13}(\lambda_1)P_{12})}_{R_{13}(\lambda_1)}\underbrace{P_{12})\hat{R}_{12}(\lambda_2)}_{R_{12}(\lambda_2)}\\ &= P_{13}R_{23}(\mu)R_{13}(\lambda_1)R_{12}(\lambda_2)\\ &= P_{13}\cdot(\text{right hand side of the (YBE)}).\end{aligned}$$

□

Remark. *If we consider parameter-independent Yang–Baxter equation, i.e., without the spectral parameters λ_1, λ_2, we get the following correspondence:* $R \in \mathrm{End}(V \otimes V)$ *solves* $R_{12}R_{13}R_{23} = R_{23}R_{13}R_{12}$ *then* $\hat{R} := PR$ *solves*

$$\hat{R}_{23}\hat{R}_{12}\hat{R}_{23} = \hat{R}_{12}\hat{R}_{23}\hat{R}_{12},$$

compare with the Artin's braid group relation.

8.3 Classical Yang–Baxter Equation (CYBE)

By considering the classical limit of the Yang–Baxter equation, i.e., by taking the first order expansion with respect to $\hbar$:

$$R(u) = 1 + i\hbar r(u) + \mathcal{O}(\hbar^2), \quad u \in \mathbb{C}$$

and by plugging

$$R_{12}(\mu) = 1 + i\hbar r_{12}(\mu) + \mathcal{O}(\hbar^2) \tag{8.28}$$

$$R_{13}(\lambda_1) = 1 + i\hbar r_{13}(\lambda_1) + \mathcal{O}(\hbar^2) \tag{8.29}$$

$$R_{23}(\lambda_2) = 1 + i\hbar r_{23}(\lambda_2) + \mathcal{O}(\hbar^2) \tag{8.30}$$

into the Yang–Baxter equation, we obtain:

$$\begin{aligned}L.H.S. &= (1 + i\hbar r_{12}(\mu) + \mathcal{O}(\hbar^2))(1 + i\hbar r_{13}(\lambda_1) + \mathcal{O}(\hbar^2))(1 + i\hbar r_{23}(\lambda_2) + \mathcal{O}(\hbar^2))\\ &= (i\hbar)^2(r_{12}(\mu)r_{13}(\lambda_1) + r_{12}(\mu)r_{23}(\lambda_2) + r_{13}(\lambda_1)r_{23}(\lambda_2)) + h.o.t.\\ R.H.S. &= (1 + i\hbar r_{23}(\lambda_2) + \mathcal{O}(\hbar^2))(1 + i\hbar r_{13}(\lambda_1) + \mathcal{O}(\hbar^2))(1 + i\hbar r_{12}(\mu) + \mathcal{O}(\hbar^2))\\ &= (i\hbar)^2(r_{13}(\lambda_1)r_{12}(\mu) + r_{23}(\lambda_2)r_{12}(\mu) + r_{23}(\lambda_2)r_{13}(\lambda_1)) + h.o.t.\end{aligned}$$

The above calculation yields the *classical* Yang–Baxter equation

$$[r_{12}(\mu), r_{13}(\lambda_1)] + [r_{12}(\mu), r_{23}(\lambda_2)] + [r_{13}(\lambda_1), r_{23}(\lambda_2)] = 0. \tag{8.31}$$

8.3.1 Frobenius Lie Algebras and Constant Solutions of CYBE

By following [2], we provide an explanation of how to construct solutions of classical Yang–Baxter equation on seaweed Lie algebras by combining the results derived in Chapter 4. Recall from Theorem 4.1, that seaweed meanders of the form

$$\mathcal{S}(a_1, \dots, a_n \mid (a_1, \dots, a_n)[1])$$

are connected. By combining Theorem 7.2 of [7] and the one-to-one correspondence on the number of connected components of seaweed meanders and collapsed seaweed meanders, it follows that seaweed meanders of the form

$$\mathcal{S}(a_1, \dots, a_n, 1 \mid 1, a_1, \dots, a_n)$$

represent seaweed Lie algebras

$$\mathfrak{s}(a_1, \dots, a_n, 1 \mid 1, a_1, \dots, a_n) \subset \mathfrak{gl}_n$$

of index one. This means in particular that for these seaweed Lie subalgebras there exists a non-degenerate bilinear skew-symmetric form B_f, compare with Section 7.1 on the index of a Lie algebra. Lie algebras, on which there exist nondegenerate bilinear functionals are called *Frobenius* Lie algebras. According to [2, 7. Rational Solutions, not Having a Pole at Infinity, 3), pp. 179–180], we construct a solution to

$$[r^{12}, r^{13}] + [r^{12}, r^{23}] + [r^{13}, r^{23}] = 0 \tag{8.32}$$

$$r^{21} = -r^{12} \tag{8.33}$$

from a Frobenius Lie algebra $\mathfrak{f}$, a non-degenerate functional $l \in \mathfrak{f}^*$ and an embedding $\mathfrak{f} \subset \mathfrak{g}$. To be more precise, we recall the following proposition, cf. [19], which ensures the above correspondence between non-degenerate bilinear forms on Frobenius Lie algebras and solutions of classical Yang–Baxter equation.

Proposition 8.2. *Let $\mathfrak{f}$ be a Lie algebra, $r \in \mathfrak{f} \wedge \mathfrak{f}$ a non-degenerate tensor and $B = B_r$ the corresponding bilinear form on $\mathfrak{f}$. Then, r is a classical r-matrix for a triangular Lie bialgebra structure on $\mathfrak{g}$, in other words r solves the classical Yang–Baxter equation, if and only if B is a 2-cocycle on $\mathfrak{f}$, i.e., for any x, y, z we have:*

$$B([x, y], z) + B([y, z], x) + B([z, x], y) = 0.$$

For a proof of the above proposition, see [19]. By the above proposition, any pair $(B, \mathfrak{f})$, where $\mathfrak{f}$ is a Lie subalgebra of $\mathfrak{g}$ yields a solution to classical Yang–Baxter equation, i.e., a triangular bialgebra structure on $\mathfrak{g}$. From Chapter 4, we know that right, respectively left one-shifted closed meanders with the same compositions of blocks for the upper and lower half-planes, $\mathbb{E}^+$ and $\mathbb{E}^-$, are connected. If we restrict ourselves to closed meanders formed by compositions of rainbow blocks in $\mathbb{E}^+$ and $\mathbb{E}^-$, we get connected collapsed seaweed meanders after applying the right, respectively left one-shift. We denote these connected collapsed seaweed meanders by

$$\mathcal{F}(a_1, \dots, a_n \mid (a_1, \dots, a_n)[1]) = \mathcal{F}^{coll}(2a_1, \dots 2a_n, 1 \mid 1, 2a_1, \dots, 2a_n)$$

in the case of the right one-shift, and by

$$\mathcal{F}(a_1, \dots, a_n \mid (a_1, \dots, a_n)[-1]) = \mathcal{F}^{coll}(1, 2a_1, \dots 2a_n \mid 2a_1, \dots, 2a_n, 1)$$

in the case of left one-shift. By Theorem 7.2 of [7] these connected collapsed seaweed meanders represent seaweed Lie algebras

$$\mathfrak{f}(2a_1, \dots 2a_n, 1 \mid 1, 2a_1, \dots, 2a_n), \text{ and } \mathfrak{f}(1, 2a_1, \dots 2a_n \mid 2a_1, \dots, 2a_n, 1),$$

which are of index 1 since the underlying collapsed seaweed meanders are connected. In particular,

$$\mathfrak{f}(2a_1, \dots 2a_n, 1 \mid 1, 2a_1, \dots, 2a_n) \text{ amd } \mathfrak{f}(1, 2a_1, \dots 2a_n \mid 2a_1, \dots, 2a_n, 1)$$

are Frobenius. By the above proposition, we obtain solutions to classical Yang–Baxter equation from

$$(B', \mathfrak{f}(2a_1, \dots 2a_n, 1 \mid 1, 2a_1, \dots, 2a_n)),$$

respectively $(B'', \mathfrak{f}(1, 2a_1, \dots 2a_n, \mid 2a_1, \dots, 2a_n, 1))$, where B' and B'' are non-degenerate bilinear forms on the corresponding Frobenius Lie algebra. In this setting, we recall the explicit form of constant solutions of the classical Yang–Baxter equation from [16]. Let us define

$$x_i \wedge x_j := \frac{1}{2}(x_i \otimes x_j - x_j \otimes x_i), \tag{8.34}$$

cf. [16]. If B_{ij} is the matrix of the bilinear form B relative to some basis $x_1, \dots, x_d$ of $\mathfrak{f}$, then the constant solution of the classical Yang–Baxter equation is given by:

$$r = \sum_{i,j} (B^{-1})_{ij} x_i \wedge x_j \in \mathfrak{f} \wedge \mathfrak{f}. \tag{8.35}$$

As was already mentioned before, by combining the results on the index of seaweed Lie algebras by A. Kirillov and V. Dergachev with our results on the right, respectively left one-shifts when restricted to meanders with rainbow blocks, we get a wide class of Frobenius Lie algebras. By the above form of constant solutions in Equation (8.35) we get in particular constant solutions to classical Yang–Baxter equation on Frobenius Lie algebras $\mathfrak{f}(2a_1, \dots 2a_n, 1 \mid 1, 2a_1, \dots, 2a_n)$ and $\mathfrak{f}(1, 2a_1, \dots 2a_n \mid 2a_1, \dots, 2a_n, 1)$.

Summary in German (Zusammenfassung)

In dieser Arbeit studieren wir Mäander mit den folgenden drei Schwerpunkten. Der erste Schwerpunkt liegt im Studium des Zusammenhangs der Mäander begründet. Hierbei entwickeln wir ein Konzept, namens *Kollaps*. Mithilfe des Kollapses beweisen wir, dass es für die Anzahl der Zusammenhangskomponenten der Seetang Mäander mit mindestens vier oberen Blöcken keine einfache geschlossene Formel ausgedrückt mithilfe des größten gemeinsamen Teilers und homogenen Polynomen vom beliebigen Grad geben kann. Wir verallgenmeinern den Kollaps für Mäander und definieren Gabeln und Kammern mithilfe derer wir hinreichende Bedingung für den *Eins-Zusammenhang* der Mäander herleiten.

Der zweite Schwerpunkt ist die kombinatorische Klassifikation der Morse Mäander, welche in der Beschreibung der Attraktoren der skalaren parabolischen partiellen Differentialgleichungen mit Neumann Randwertbedingungen wichtig sind und auf frühere Arbeiten von B. Fiedler, C. Rocha, G. Fusco zurückgehen. Mithilfe der Links- und Rechts-Eins-Verschiebung, sowie unter Benutzung zwei weiterer Abbildungen, ist es möglich vier verschiedene Typen der Morse Mäander zu konstruieren. Entsprechend, nennen wir die vier verschiedenen Typen der Morse Mäander *Typ I, II, III, IV*.

Der dritte Schwerpunkt verfolgt die Fragestellungen aus B. Fiedlers Arbeit in [9]. Wir studieren die gerichteten Verbindungsgraphen, die mittels Permutationen der Morse Mäander vom Typ I, II, III, IV bestimmt sind, im Hinblick auf die Graphisomorphie. Hierbei, beantworten wir, wann zwei Verbindungsgraphen vom Typ I isomorph sind und darüberhinaus beweisen wir, dass zwei Verbindungsgraphen vom unterschiedlichen Typ als gerichtete Graphen stets *nicht* isomorph sind.

Zum Schluss geben wir einen Einblick in bekannte Beziehungen der Seetang Mäander zu den Seetang Lie Algebren und deren Indexberechnung nach A. Kirillov und V. Dergachev und ordnen unsere Ergebnisse bezüglich des Kollapses in diesen Bereich ein. Weiterhin, wiederholen wir die Bedeutung und den Ursprung der Yang-Baxter Gleichung und ihrer klassischen Form und integrieren unsere Ergebnisse zu den Morse Mäandern in die bereits nach A. Belavin und V. Drinfel'd bekannte Konstruktion konstanter Lösungen der klassischen Yang–Baxter Gleichung auf den Frobenius Lie Algebren.

DOI 10.1515/9783110533026-010

„Liebe deine Geschichte.
Es ist der Weg, den Gott mit dir gegangen ist."

Leo Tolstoi

Bibliography

[1] V. I. Arnol'd. A branched covering of $CP^2 \to S^4$, hyperbolicity and projectivity topology. *Sib. Math. J.*, 29(5):717–726, 1988.

[2] A. A. Belavin and V. G. Drinfel'd. Solutions of the classical Yang–Baxter equation for simple Lie algebras. *Functional Analysis and Its Applications*, 16:159–180, 1982.

[3] P. Brunovsky and B. Fiedler. Connecting orbits in scalar reaction diffusion equations. *Dynamics Reported*, 1:57–89, 1988.

[4] P. Brunovsky and B. Fiedler. Connecting orbits in scalar reaction diffusion equations II: The complete solution. *J. Differential Equations*, 81:106–135, 1989.

[5] V. Coll, M. Hyatt, C. Magnant, and H. Wang. Meander graphs and Frobenius seaweed Lie algebras. II. *J. Gen. Lie Theory Appl.*, 9(1):7, 2015.

[6] V. Coll, C. Magnant, and H. Wang. The signature of a meander. arXiv:1206.2705, 2012.

[7] V. Dergachev and A. Kirillov. Index of Lie algebras of seaweed type. *J. Lie Theory*, 10(2):331–343, 2000.

[8] P. Di Francesco, O. Golinelli, and E. Guitter. Meanders and the Temperley–Lieb algebra. *Commun. Math. Phys.*, 186(1):1–59, 1997.

[9] B. Fiedler. Global attractors of one–dimensional parabolic equations: sixteen examples. *Tatra Mountains Mathematical Publications*, 4:67–92, 1994.

[10] B. Fiedler and P. Castañeda. Rainbow meanders and Cartesian billiards. *São Paulo J. Math. Sci.*, 6(2):247–275, 2012.

[11] B. Fiedler and C. Rocha. Heteroclinic orbits of semilinear parabolic equations. *J. Differ. Equations*, 125(1):239–281, 1996.

[12] B. Fiedler and C. Rocha. Realization of meander permutations by boundary value problems. *J. Differ. Equations*, 156(2):282–308, 1999.

[13] B. Fiedler and C. Rocha. Connectivity and design of planar global attractors of Sturm type. III: Small and platonic examples. *J. Dyn. Differ. Equations*, 22(2):121–162, 2010.

[14] B. Fiedler and C. Rocha. Nonlinear Sturm global attractors: unstable manifold decompositions as regular CW-complexes. *Discrete Continuous Dynamical Systems.*, 34(12):5099–5122, 2014.

[15] G. Fusco and C. Rocha. A permutation related to the dynamics of a scalar parabolic PDE. *J. Differ. Equations*, 91:75–94, 1991.

[16] M. Gerstenhaber and A. Giaquinto. Boundary solutions of the classical Yang–Baxter equation. *Lett. Math. Phys.*, 40(4):337–353, 1997.

[17] D. Henry. Geometric theory of semilinear parabolic equations. *Lect. Notes Math.*, 840, 1981.

[18] A. Karnauhova and S. Liebscher. Connected components of meanders: I. Bi–rainbow meanders. To appear, 2015.

[19] L. I. Korogodski and Y. S. Soibelman. Algebras of functions on quantum groups: Part 1. *American Mathematical Society*, 56:150, 1998.

[20] D. Kozlov. *Combinatorial algebraic topology.* Springer Verlag New York Berlin Heidelberg, 2008.

[21] M. La Croix. Approaches to the enumerative theory of meanders. Master's thesis, www.math.uwaterloo.ca/~malacroi/Latex/Meanders.pdf, 2003.

[22] O. A. Ladyzhenskaya. Attractors for semi–groups and evolution equations. *Cambridge University Press*, 1991.

[23] H. Matano. Convergence of solutions of one–dimensional semilinear parabolic equations. *J. Math. Kyoto Univ.*, 18:221–227, 1978.

[24] D. I. Panyushev. Inductive formulas for the index of seaweed Lie algebras. *Moscow Mathematical Journal*, 1(2):221–241, 2001.

[25] C. Rocha. Properties of the attractor of a scalar parabolic PDE. *Journal of Dynamics and Differential Equations*, 3(4):557–591, 1991.

[26] L. Samaj and Z. Bajnok. Introduction to statistical physics of integrable many-body systems. *Cambridge University Press*, 2013.

[27] H. Wenzl. Representations of braid groups and the quantum Yang–Baxter equation. *Pacific J. of Math.*, 145(1):221–227, 1990.

[28] J. H. C. Whitehead. Combinatorial homotopy. I. *Bull. Am. Math. Soc.*, 55:213–245, 1949.

[29] M. Wolfrum. A sequence of order relations: encoding heteroclinic connections in scalar parabolic PDE. *Journal of Differential Equations*, 183:56–78, 2002.

[30] C. N. Yang. Some exact results for the many–body problem in one dimension with repulsive delta–function interaction. *Phys. Rev. Lett.*, 19:1312–1315, 1967.

[31] E. Zeidler. *Nonlinear functional analysis and its applications II/B: Nonlinear Monotone Operators.* Springer Verlag New York Berlin Heidelberg, 1990.

[32] T. I. Zelenyak. Stabilization of solutions of boundary value problems for a second order parabolic equation with one space variable. *J. Differential Equations*, 4:17–22, 1968.

Bei Fragen zur Produktsicherheit wenden Sie sich bitte an:
If you have any questions regarding product safety,
please contact:

Walter de Gruyter GmbH
Genthiner Straße 13
10785 Berlin
productsafety@degruyterbrill.com